PHARMACEUTICAL MANUFACTURERS

PHARMACEUTICAL MANUFACTURERS

An International Directory

by

David B. Braun

np
NOYES PUBLICATIONS
Park Ridge, New Jersey, U.S.A.

Library of Congress Catalog Card Number: 95-21383
ISBN: 0-8155-1384-4
Printed in the United States

Published in the United States of America by
Noyes Publications
Mill Road, Park Ridge, New Jersey 07656

10 9 8 7 6 5 4 3 2 1

Library of Congress Cataloging-in-Publication Data

Braun, David B.
Pharmaceutical manufacturers : an international directory / by David B. Braun.
p. cm.
ISBN 0-8155-1384-4
1. Pharmaceutical industry--Directories. 2. Drug packaging industry--Directories. I. Title.
HD9665.3.B73 1995
615.1'029'4--dc20 95-21383
CIP

PREFACE

This directory contains the name and address of more than 1,050 pharmaceutical manufacturers in 50 nations around the world. They are arranged alphabetically by country and company name.

Most entries also include a telephone and facsimile (FAX) number for the company's location. The user is advised that these numbers do not include the country code number. There are found in your local telephone directory. The city code is almost always provided. The most notable exceptions are the developing nations such as China where city code numbers are nonexistent or not readily available.

The names and titles of one or more key personnel at the location are also incorporated in the entry, when available.

Completing the directory, is a listing of 124 Contract Manufacturers of ethical and/or non-prescription pharmaceuticals. These are arranged alphabetically by country and company name.

The pharmaceutical industry is very dynamic and the reader may find that some manufacturers listed have merged or been "acquired" by other companies since this information was compiled. The data contained herein is believed to be accurate as of the Spring of 1995.

CONTENTS

PART I
PHARMACEUTICAL MANUFACTURERS

PART II
CONTRACT PACKAGERS OF PHARMACEUTICALS

PART I

PHARMACEUTICAL MANUFACTURERS

1. ARGENTINA

Abbott Laboratorios Argentina S.A.
Ruta 2 Km 34, Florencio Varela
Buenos Aires 1888

Telephone: 1 258 8860

Key Personnel, This Location:
Ricardo P. Storm, President

Parent Company:
Abbott Laboratories

Bernardo y Cia. S.A.C.I.F.
Terrada 2346
Buenos Aires 1416

Telephone: 1 501 3728

Key Personnel, This Location:
Ariel A. Liebstein Solia, Manager

Byk Liprandi S.A.C.I.
Delgado 1549
Buenos Aires 1426

Telephone: 1 552 3346

Parent Company:
Byk Gulden Lomberg, GmbH

Chemotechnia Sintyal S.A.P.O.I.F.
Carlos Berg 3669
Buenos Aires 1437

Telephone: 1 777 2892

Ciba-Geigy Argentina S.A.I.C.F.
Arias 1851
Buenos Aires 1429

Parent Company:
Ciba-Geigy AG

Telephone: 1 701 2461

Key Personnel, This Location:
Alfonso Cristobal Freyvogel

Drogueria Fuchs, S.A.I.C.F.
Moreno 1833
Buenos Aires 1094

Telephone: 1 476 4581

Key Personnel, This Location:
Maria Carmen Balo Fuchs, President

Essex Argentina S.A.I.C.
Av. San Martin 4550
Los Polvorines, Buenos Aires 1752

Parent Company:
Schering Overseas Ltd.

Telephone: 1 652 7334

Finadiet S.A.C.I.F.
H. Yrigoyen 3771
Buenos Aires 1208

Instituto Sidus I.C.S.A.
Larrea 926
Buenos Aires 1117

Telephone: 1 180 2700

Kolynos S.A.C.I.
Ty Gral J D Peron 925
Buenos Aires 1038

Telephone: 1 250 1010

Parent Company:
Colgate-Palmolive Corp.

Laboratorio Elea S.A.C.I.F.A.
Saladillo 2540
Buenos Aires 1440

Laboratorios Andromaco S.A.I.C.I
Ing Huergo 1139
Buenos Aires 1107

Telephone: 1 829 6274

Laboratorios Bago S.A.
Bernardo De Yrigoyen 248
Buenos Aires 1072

Telephone: 1 224 5224

Laboratorios Beta S.A.
San Juan 2266
Buenos Aires 1232

Telephone: 1 941 1001

Laboratorios Casasco S.A.I.C.
Boyaca 237
Buenos Aires 1406

Telephone: 1 632 5091

Key Personnel, This Location:
E.F. Casasco, President

Laboratorios Dr. Gador y Cia. S.A.C.I.
Florida 868
Buenos Aires 1005

Telephone: 1 226 3542

Laboratorios Glaxo Argentina S.A.C.I.
J. J. Castelli 6701
Buenos Aires 1605

Telephone: 1 766 0021

Parent Company:
Glaxo Group, plc

Laboratorios Promeco S.A.C.I.F.
Chile 80
Buenos Aires 1098

Telephone: 1 255 1000

Laboratorios Temis Lostalo S.A.
Zepita 3178
Buenos Aires 1285

Telephone: 1 285 000

Laplex S.A.
San Martin 910, Piso 4
Buenos Aires

Telephone: 1 322 5920

Key Personnel, This Location:
Julio C. Guardone, President

Lepetit S.A.
Av. Leanadro N. Alem 896
Buenos Aires 1001

Telephone: 1 220 3257

Merck Sharp y Dohme Argentina, Inc.
Av. del Liberatador 1400,
Victoria, Buenos Aires 1638

Telephone: 1 797 2071

Newark Especialidades Medicinales
Av. Federico Lacroze 3065
Buenos Aires

Telephone: 1 772 0779

Key Personnel, This Location:
Hector Rodrigue Acosta, Co-owner

Pinal Pharma S.A.C.I.F.
Juan B. Alberdi 581
Buenos Aires 1424

Telephone: 1 99 1188

Key Personnel, This Location:
J.C. Biedma, President

Quimica Estrella S.A.C.I.I.
Avienda De Los Consituyentes 29
Buenos Aires 1427

Telephone: 1 775 4037

Quimica Industrial Mebomar S.A.
General Urquiza
Buenos Aires 1164

Telephone: 1 939 6520

Key Personnel, This Location:
Hector A. Monzon, President

Quimica Montpellier S.A.
Virrey Liniers 667
Buenos Aires 1220

Telephone: 1 224 5622

Gerardo Ramon & Cia. S.A.I.C.
Int Cnel Amaro Avalos 4208
Buenos Aires 1605

Telephone: 1 488 0910

Key Personnel, This Location:
G. S. L. Ramon, President

Riedel y Lavalle S.R.L.
Dr. Juan F. Aranguren 2955
Buenos Aires

Telephone: 1 611 7480

Key Personnel, This Location:
Lazaro Szlukier, Co-owner

Rhône-Poulenc Argentina S.A.
Primera Junta 525,
Ramello, Buenos Aires 1878

Telephone: 1 225 0372

Roemmers S.A.I.C.F.
Corrientes 316
Buenos Aires 1100

Telephone: 1 38 8131

Key Personnel, This Location:
J. R. F. Roemmer, Director

Syncro Argentina S.A.Q.I.C.I.F.
Sarmiento 1230
Buenos Aires 1041

Telephone: 1 35 5475

Szabo Hnos Kessler & Cia S.R.L.
Humahuaca 1065-79
Buenos Aires

Telephone: 1 868 436

Key Personnel, This Location:
Gerardo Kessler, Manager

---

2. AUSTRALIA

C S L, Ltd.
45 Poplar Road
Parkville, Victoria 3052

Telephone: 3 389 1911 FAX: 3 389 1434

Key Personnel, This Location:
John Howlett, Marketing Manager

Essex Laboratories Pty., Ltd.
P.O. Box 231
Sydney, New South Wales 2153

Telephone: 2 624 4444

Key Personnel, This Location:
K.R. Wilson, Managing Director

F. H. Faulding & Company, Ltd.
160 Greenhill Road
Parkside, South Australia 5063

Telephone: 8 372 1500

Glaxo Australia P/L
1061 Mountain Highway
Boronia, Victoria 3155

Telephone: 3 729 5100

Parent Company:
Glaxo Group, plc

Kiwi Brands P/L
610 Heatherton Road
Clayton, Victoria 3168

Telephone: 3 262 2300

Parent Company:
Sara Lee Holdings Australia

Eli Lilly Australia Pty., Ltd.
112 Wharf Road
Sydney, New South Wales 2114

Telephone: 2 858 8022 FAX: 2 874 1613

Key Personnel, This Location:
Douglas C. Vawter

3M Pharmaceuticals Pty., Ltd.
9-15 Chilvers Road
Thornleigh, New South Wales 2120

Telephone: 2 875 6333 FAX: 2 875 6416

Key Personnel, This Location:
N. Mills, Director

Merck Sharp & Dohme Australia
P.O. Box 79
Sydney, New South Wales 2142

Telephone: 2 645 0555 FAX: 2 645 1369

Key Personnel, This Location:
A.J. Graham, Managing Director

Pharmaction Pty., Ltd.
17-19 John Street
Brunswick East, Victoria 3157

Telephone: 3 388 0911 FAX: 3 387 2668

Key Personnel, This Location:
Suzanne Aenall, Managing Director

R & C Products P/L
33 Hope Street
Ermington, New South Wales 2115

Parent Company:
Reckitt & Colman Austrailia, Ltd.

Telephone: 2 858 8111

A.H. Robins Pty., Ltd.
Private Bag 1
Sydney, New South Wales 2196

Parent Company:
American Home Products Corp.

Telephone: 2 534 1000 FAX: 2 533 3980

Schering-Plough Pty., Ltd.
11 Gibbon Road
Baulkham Hills, New South Wales 2153

Telephone: 2 624 4444 FAX: 2 674 4279

Key Personnel, This Location:
David Hancock, Managing Director

Searle Australia Pty., Ltd.
20 Clarke Street
Crows Nest, New South Wales 2065

Telephone: 2 439 6788 FAX: 2 438 4930

Key Personnel, This Location:
Alan Walters

Sigma Company, Ltd.
1408 Centre Road, P.O. Box 144
Clayton, Victoria 3168

Telephone: 3 542 9511 FAX: 3 542 9769

Key Personnel, This Location:
R. Weller, General Manager

Washington H. Soul Pattinson Co., Ltd.
160 Pitt Street Mall
Sydney, New South Wales 2000

Telephone: 2 232 7166 FAX: 2 235 1747

Key Personnel, This Location:
P. Robinson, C.E.O.

---

3. AUSTRIA

Abbott GmbH
Diefenbachgasse 35
1150 Vienna

Telephone: (1) 891 220

Parent Company:
Abbott Laboratories

Aesca chem.-pharm. Fabrika GmbH
Badener Strasse 23
2514 Traiskirchen

Telephone: (2252) 522 870 FAX: (2252) 542 94

Albert-Roussel Pharma GmbH
Altsmanndorfer Strasse 104
1121 Vienna

Telephone: (1) 801 010 FAX:(1) 801 01 28 08

Apomedica GmbH & Company KG
Roseggerkai 3
8011 Graz

Telephone: (316) 823 53 30 FAX: (316) 823 53 352

Astra GmbH
Südtirolerstrasse 6
4020 Linz

Telephone: (732) 66 72 310 FAX: (732) 66 84 21

Parent Company:
Astra AB

Austroplant-Arzneimittel GmbH
Richard Strauss Strasse 13
1232 Vienna

Telephone: (1) 616 26 440 FAX: (1) 616 26 44 18

Bayer Austria GmbH
Am Heutmarkt 10
1037 Vienna

Parent Company:
Bayer AG

Telephone: (1) 711 460 FAX: (1) 711 46 14

Bender & Company GmbH
Dr. Boehringer-Gasse 5-11
1121 Vienna

Telephone: (1) 801 050 FAX: (1) 804 08 23

Biochemie GmbH
Dr. Hans Bachmannstrasse 7
6250 Kundl, Tirol

Parent Company:
Sandoz AG

Telephone: (5338) 200 0 FAX: (5338) 200 46

Key Personnel, This Location:
Heinrich Scherfler, Dipl. Ing.

Boehringer Mannheim GmbH
Engelhorngasse 3
1211 Vienna

Telephone: (1) 277 87 0 FAX: (1) 277 87 12

Byk Österreich Pharma GmbH
Ketzergasse 200
1235 Vienna

Telephone: (1) 860 486 FAX: (1) 860 486 40

Ciba-Geigy GmbH
Breitenfurter Strasse 251
1231 Vienna

Parent Company:
Ciba-Geigy AG

Telephone: (1) 801 11 FAX: (1) 801 11 466

Cilag GmbH
Pfarrgasse 75
1232 Vienna

Telephone: (1) 616 12 50 FAX: (1) 616 12 41

EBEWE Arzneimittel GmbH
Mondseestrasse 11
4866 Unterach

Telephone: (7665) 81 230 FAX: (7665) 81 23 11

Germania Pharmazeutika GmbH
Schuselkagasse 8
1150 Vienna

Telephone: (1) 982 33 99 FAX: (1) 982 33 99 24

Gerot Pharmazeutika GmbH
Arnethgasse 3
1171 Vienna

Telephone: (1) 435 50 50 FAX: (1) 409 29 21

Gödecke GmbH
Ketzergasse 118
1234 Vienna

Telephone: (1) 86 24 150 FAX: (1) 86 24 154

Hoechst Austria AG
Altmannsdorfer Strasse 104
1121 Vienna

Parent Company:
Hoechst AG

Telephone: (1) 801 010 FAX: (1) 804 22 08

Hoffman-LaRoche Wein GmbH
Jacquingasse 16-18
1030 Vienna

Parent Company:
Hoffman-LaRoche, Inc.

Telephone: (1) 781 604 FAX: (1) 799 124 4

Jacoby Pharmazeutika GmbH
Teighweg 2
5400 Hallein

Telephone: (6245) 255 10 FAX: (6245) 255 168

Kabi Pharmacia GmbH
Oberlaaer Strasse 251
1100 Vienna

Telephone: (1) 686 638 FAX: (1) 686 63 84

Kali-Pharma GmbH
Donaustrasse 106
3400 Klosterneuburg

Telephone: (2243) 856 20 FAX: (2243) 856 20 40

Klinge Pharma GmbH
Hietzinger Hauptstrasse 64
1132 Vienna

Telephone: (1) 877 266 8 FAX: (1) 877 16 36

Laevosan GmbH
Estermannstrasse 17
4020 Linz

Telephone: (732) 76 510 FAX: (732) 78 28 33

Lannacher Heilmittel GmbH
Schlossplatz 1
8502 Lannach

Telephone: (3136) 825 77 FAX: (3136) 815 63

Leopold Pharma GmbH
Hafnerstrasse 36
8055 Graz-Puntigam

Telephone: (316) 29 55 50 FAX: (316) 29 55 50 269

Madaus GmbH
Lienfeldergassse 91-93
1171 Vienna

Telephone: (1) 450 16 00 0 FAX: (1) 456 16 00 25

Mayhofer Pharmazeutika GmbH
Herrenstrasse 2
4010 Linz

Telephone: (732) 76 590 FAX: (732) 76 594

Mundipharma GmbH
Apollogasse 16-18
1072 Vienna

Telephone: (1) 523 25 05 FAX: (1) 523 25 0544

Parke-Davis Gmbh
Ketzergasse 118
1234 Vienna

Parent Company:
Warner-Lambert Company

Telephone: (1) 86 24 150 FAX: (1) 86 24 1543

Pfizer Corp. Austria GmbH
Mondscheingasse 16
1070 Vienna

Parent Company:
Pfizer, Inc.

Telephone: (1) 521 15-0 FAX: (1) 526 91 32

Pharmazeutische Fabrik Montavit GmbH
Salzbergstrasse 96
6060 Hall/Tirol

Telephone: (5223) 7926 FAX: (5223) 79 26 11

Röhm Pharma GmbH
Tendlergasse 13
1090 Vienna

Telephone: (1) 43 72 55 FAX: (1) 402 96 91 22

Sandoz GmbH
Brunner Strasse 59
1235 Vienna

Parent Company:
Sandoz AG

Telephone: (1) 86 45 460 FAX: (1) 86 45 46 721

Waldheim Pharmazeutika GmbH
Boltzmanngasse 11
1090 Vienna

Telephone: (1) 319 14 560 FAX: (1) 319 14 56 44

Wellcome Austria GmbH
Inzersdorfer Strasse 64
1100 Vienna

Parent Company:
The Wellcome Foundation, Ltd.

Telephone: (1) 604 88 880 FAX: (1) 604 88 881

Zeneca Österreich GmbH
Schwarzenbergplatz 7
1037 Vienna

Telephone: (1) 711 310 FAX: (1) 711 312 21

---

4. BELGIUM

Abbott Laboratories
Parc Scientifique
Rue du Bosquet 2
1348 Louvain-la-Neuve

Parent Company:
Abbott Laboratories

Telephone: 10 475 311

FAX: 10 475 575

Key Personnel, This Location:
Veronique Mees

Astra Pharma
Chaussee d'Alsemberg 1001
1180 Brussels

Parent Company:
Astra AB

Telephone: 2 370 4611

FAX: 2 370 4688

Key Personnel, This Location:
Ms. Ven Den Bossche

Bristol Myers Squibb Belgium
Chaussee de la Hulpe 185
1170 Brussels

Parent Company:
Bristol Myers Squibb Company

Telephone: 2 674 0411

FAX: 2 674 0600

Key Personnel, This Location:
Wilfred Van Loo

Omnichem S.A.
Rue Fonds Jean Paques 8
Mont-Saint-Gulbert, 1435 Brabant

Telephone: 2 045 0031

Pfizer-Roerig
Rue L. Theodore 102
1090 Brussels

Parent Company:
Pfizer, Inc.

Telephone: 2 423 0511 FAX: 2 423 0799

Key Personnel, This Location:
D. DeGroode

Purna Pharmaceuticals
Rijksweg 17
2870 Puurs

Telephone: 3 886 0085 FAX: 3 886 2538

Key Personnel, This Location:
Mr. Van Gucht

Qualiphar
Rijkweg 9
2880 Bornem

Telephone: 3 889 1721 FAX: 3 889 6913

Key Personnel, This Location:
Mr. Van Eck

Sanofi-Pharma S.A.
Avenue De Bejar 1Elf
Bruxelles, Barbant 1120

Telephone: 2 266 4311

Parent Company:
Elf Sanofi S.A.

SmithKline Beecham Pharma
Rue Du Tilleul 13
1332 Genval

Telephone: 2 656 2111

FAX: 2 656 2499

Parent Company:
SmithKline Beecham plc

Key Personnel, This Location:
Mark Januszewski, Director
M. De Winter

Upjohn NV
Rijksweg 12
2870 Puurs

Telephone: 3 890 9211

FAX: 3 889 8475

Parent Company:
The Upjohn Co.

Key Personnel, This Location:
Muriel D'Hondt

---

5. BRAZIL

Alcon Laboratorios do Brasil S.A.
Av. Nossa Senhora da Assuncao 736
Sao Paulo, SP 05350-001

Telephone 11 268 7433 FAX: 11 819 2171

Abbott Laboratorios Do Brasil Ltda.
Rua Nova York 245 Brooklin
Sao Paulo, SP 04560

Parent Company:
Abbott Laboratories

Telephone: 11 240 13 22

Anakol Industria e Comercio Ltda.
Rua Caetano Ointo 129 - 3 Andar
Sao Paulo, SP 03041

Telephone: 11 278 88 22 FAX: 11 278 61 03

Key Personnel, This Location:
Jose E. Franze, Director

BASF Brasileira S.A.
Av. Sao Luiz, Terreo 1 a 21 a
Sao Paulo, 01046

Telephone: 11 545 2181

Key Personnel, This Location:
Heinz W.H. Wollenweber, Director President

Casa Granado Laboratorios Farmacias
Rua Primeiro de Marco 14/16
Rio de Janeiro, RJ 23010

Telephone: 21 224 56 28

Fina Quimica Farmaceutica S.A.
Rua Frederico Rene de Jaegher 268
Sao Paulo, SP 04826

Telephone: 11 520 97 11 FAX: 11 520 62 29

Key Personnel, This Location:
Flavio Alves da Rocha, President

Industria Quimica do Estado de Goias S.A.
Avenida Anhanguera 9.827
Goiania, GO 74000

Telephone: 62 271 29 11

Key Personnel, This Location:
Jovair O. Arantes, President

Industria Quimica E Farmaceutica Schering-Plough S.A.
Est Dos Bandeirantes 3091 C Deus
Rio De Janeiro, RJ 22775

Parent Company:
Schering Overseas, Ltd.

Telephone: 21 342 70 00

Johnson & Johnson S.A.
Rua Gerivativa 207 Butanta
Sao Paulo, SP 05501

Parent Company:
Johnson & Johnson

Telephone: 11 210 50 00

FAX: 11 813 78 42

Key Personnel, This Location:
C. A. Gottschalk, President

Laboratorios Brusch Ltda.
Rua Alfredo Pereina Sampaio 120
Sao Paulo, SP 04676-011

Telephone: 11 246 1194

FAX: 11 524 8186

Key Personnel, This Location:
Matilde Guembes, Technical Director

Laboratorios Wyeth Ltda.
Via Anchieta S/N Km 14
Rudge Ram
Sao Bernardo Do Campo, SP 09883

Parent Company:
American Home Products Corp.

Telephone: 11 457 88 22

Leivas Leite S.A.
Rua Benjamin Constant 1537
Pelotas, RS 96010-020

Telephone: 532 25 3666

3M do Brazil Ltda.
Via Anhanguera Km. 110
Sumare, SP 13181

Telephone: 192 641 765 FAX: 192 641 206

Key Personnel, This Location:
Jacintho Manara, Supervisor

Rhodia S.A.
Av. Maria Coelho de Aguiar 215
Sao Paulo, SP 05805

Parent Company:
Rhône-Poulenc S.A.

Telephone: 11 545 77 62 FAX: 11 548 18 05

Key Personnel, This Location:
Jean Martin Sigrist, President

Uniao Quimica Farmaceutica
Av. dos Bandeirantes 5386
Sao Paulo, SP 04071-001

Telephone: 11 196 2100

Key Personnel, This Location:
Joao M. dePaulo, President

---

6. CANADA

Abbott Laboratories, Ltd.
6300 Cote De Liesse Ch
Montreal, Québec 3C3 K6

Telephone: (514) 340-7100

Key Personnel, This Location:
R.W. Hooper, President

Parent Company:
Abbott Laboratories

Apotex Inc.
150 Signet Drive
Weston, Ontario M9L 1T7

Telephone: (416) 513-5800

Astra Pharma, Inc.
1004 Middlegate Road
Mississauga, Ontario 4Y1 M4

Telephone: (416) 275-7111

Parent Company:
Astra AB

Ayerst McKenna & Harrison, Inc.
1025 Boulevard Laurentian
Montreal, Québec H4R 1J

Telephone: (514) 744-6771

Parent Company:
American Home Products Corp.

Baxter Corp.
6695 Airport Road
Mississauga, Ontario L4V 1T7

Telephone: (416) 673-2255

Parent Company:
Baxter International, Inc.

Beecham Laboratories, Inc.
115 Boul. Brunswick
Point Clair Dor, Québec H9R 1A4

Telephone: (514) 695-6551

Parent Company:
Beecham Canada, Inc.

Block Drug Company (Canada)
7600 Danbro Creasant
Mississauga, Ontario 5N6 L6

Telephone: (416) 542-7282

Parent Company:
Block Drug Co.

Boehringer Ingleheim (Canada), Ltd.
977 Century Drive
Burlington, Ontario L7L 5J8

Telephone: (416) 639-0333

Burroughs Wellcome, Inc.
16751 Rte Trans-Canadian
Kirkland, Québec H9H 4J4

Telephone: (514) 694-8220 FAX: (514) 694 8201

Key Personnel, This Location:
Danielle Munn, Manager

Confab Laboratories, Inc.
4355 Sir Wilfred Laurier
Sainte Hubert, Québec J3Y 3X3

Telephone: (905) 737 9200 FAX: (905) 737 8281

Eli Lilly Canada, Inc.
3650 Danforth Avenue
Scarborough, Ontario M1N 2E8

Telephone: (416) 694-3221

Parent Company:
Eli Lilly & Company

Glaxo Canada, Inc.
7333 Mississauga Road North
Mississauga, Ontario L5n 6L4

Telephone: (905) 819-3000 FAX: (905) 819-3099

Key Personnel, This Location:
S. Gilmer, Manager

Hoechst Canada, Inc.
3655 Côte d'En Bas
Varennes, Québec J3X 1T6

Telephone: (514) 652 2946 FAX: (514) 652 5200

Key Personnel, This Location:
Tosh Oike, Plant Manager

Frank W. Horner, Inc.
8485 Ferrier
Montreal, Quebec H4P 1M6

Telephone: (514) 731-3931

Key Personnel, This Location:
J. Robert Fraser, President
R. Sowatzky, Vice President

Hyal Pharmaceutical Corp.
3909 Nashua Drive, Unit 5
Mississagua, Ont L4V 1R3

Telephone: (416) 678-6800

Key Personnel, This Location:
S.S. Ascula, President
M.A. Byrne, Vice President

ICN Canada, Ltd.
1956 Bourdon Rue
Montreal, Québec 4M1 1V1

Johnson & Johnson, Inc.
2155 Pie Lx
Montreal, Québec H1V E4

Telephone: (514) 251-5151

Parent Company:
Johnson & Johnson

McNeil Pharmaceutical
600 Main Street
Stouffville, Ontario 4A7 X7

Telephone: (416) 640-6900

Parent Company:
Johnson & Johnson

Merck Frosst Canada, Inc.
16711 Route Trans-Canadian
Roxboro, Québec H9H 3L1

Telephone: (514) 695-7920

Merrell Dow Pharmaceuticals, Ltd.
380 Elgin Mills Road
Richmond Hill, Ontario 4C5 H2

Telephone: (416) 883-1915

Novopharm, Ltd.
30 Nably Court
Scarborough, Ontario M1B 2K9

Telephone: (416) 291-8876

Organon Canada, Ltd.
565 Coronation Drive
West Hill, Ontario M1E 2K3

Telephone: (416) 284-6131

Ortho Pharmaceutical (Canada)
19 Greenbelt Drive
Don Mills, Ontario M3C 1L9

Telephone: (416) 449-9444

Parent Company:
Johnson & Johnson Intl.

Parke-Davis
2337 Parkedale Avenue
Brockville, Ontario K6V 5W5

Parent Company:
Warner-Lambert Co.

Telephone: (613) 342-4436

FAX: (613) 342-6584

Key Personnel, This Location:
Norman Reynolds, Manager, Technical Services
Ivan Zarenda, Plant Manager

Pfizer Canada, Inc.
P.O. Box 800, Point Claire
Dorval, Québec H9R 4V2

Parent Company:
Pfizer, Inc.

Telephone: (514) 426-7418

FAX: (514) 426-7013

Key Personnel, This Location:
G. Patterson, Operations Manager

Professional Pharmaceutical
9200 Ch Cote De Liesse
Lachine, Québec H8T 1A1

Parent Company:
M&P Cape Investments, Ltd

Telephone:(514) 631-7710

Purdue Frederick
575 Granite Court
Pickering, Ontario L1W 3W8

Parent Company:
Purdue Fredrick, USA

Telephone: (905) 420-6400

FAX: (905) 420-4193

Key Personnel, This Location:
J.H.F. Rider, Exec. Vice President
John Stewart, Exec. Vice President

Rhône-Poulenc Rorer Canada, Inc.
8580 Avenue De L'Esplanade
Montreal, Québec 2P2 R9

Telephone: (514) 384-8220

Parent Company:
Rhône-Poulenc Rorer, Inc.

A. H. Robins Canada, Inc.
2360 Southfield Road
Mississauga, Ontario L5N 3R6

Telephone: (416) 821-8820

Parent Company:
American Home Products Corp.

Rorer Canada, Inc.
130 East Drive
Brampton, Ontario I6T 1C3

Telephone: (416) 792-1212

Parent Company:
Rhône-Poulenc Rorer, Inc.

Schering Canada, Inc.
3535 Route Trans-Canadian
Point-Clair-Dor, Québec 9R1 B4

Telephone: (514) 694-1320

Parent Company:
Schering Corp.

Schering-Plough Healthcare
6400 Northam Drive
Mississauga, Ontario L4V 1H9

Telephone: (416) 673-6242

Parent Company:
Schering Corp.

Searle Canada, Inc.
400 Iroquois Shore Road
Oakville, Ontario L6H 1M5

Telephone: (416) 844-1040

Parent Company:
G.D. Searle & Co.

Syntex, Inc.
2100 Syntex Court
Mississauga, Ontario L5N 3X4

Telephone: (905) 821-4000

FAX: (905) 821-2902

Taro Pharmaceuticals, Inc.
130 East Drive
Bramlea, Ont L6T 1C3

Telephone: (905) 791-8276

Parent Company:
Taro Pharmaceuticals Ind., Ltd

FAX: (905) 791-5008

Therapex
11100 Colbert
Anjou, Québec H1J 2M9

Telephone: (514) 353 5820

Key Personnel, This Location:
Pierre Andre Ouimet, President

Parent Company:
E-Z-EM Corp.

Tican Pharmaceuticals (Canada), Ltd.
100 St. Joseph Boulevard, West
Montreal, Québec H2T 2P6

Telephone: (514) 271 5255

FAX: (514) 271 7861

Key Personnel, This Location:
George Phillips, Director of Marketing

Upjohn Company Of Canada
865 York Mills Road
Don Mills, Ontario M3B 1Y6

Parent Company:
The Upjohn Company

Telephone: (416) 441-9600

FAX: (416) 445-9048

Key Personnel, This Location:
Don J. Horan, Director, Business Development
Cheryl Kalpin, Manager, Export/Contract Packaging

Vita Health Co., Ltd
150 Beghin Avenue
Winnipeg, Manitoba R2J 3W2

Telephone: (204) 661-8386

Wampole, Inc.
6130 East Drive
Brampton, Ontario L6T 1C3

Telephone: (416) 792-1212

Warner-Lambert Canada, Inc.
2200 Eglington Avenue East
Scarborough, Ontario M1L 2N3

Parent Company:
Warner-Lambert Company

Telephone: (416) 288-2200

Whitehall Laboratories, Ltd.
1060 Middlegate Road
Mississauga, Ontario L4Y 1M3

Parent Company:
American Home Products Co.

Telephone: (416) 277-2702

Wyeth Ayerst Canada, Ltd.
1025 Laurentian Boulevard
Saint Laurent (Montreal) Québec H4R 1J6

Parent Company:
American Home Products Co.

Telephone: (514) 744-6771
FAX: (514) 744-3249

---

7. CHILE

Abbott Laroratories de Chile, Ltda.
Casilla 169-D
Santiago

Parent Company:
Abbott Laboratories

Telephone: 2 557 60 62

FAX: 2 557 59 63

Key Personnel, This Location:
Rodolfo Quevedo, General Manager

Farmacia Reccius S.A.
San Antonio 229
Santiago

Telephone: 2 639 21 31

FAX: 2 233 77 07

Key Personnel, This Location:
Raul Vasquez, President

Laboratorios Chile SA
Avienda Marathon 1315
Santiago, Santiago

Telephone: 2 238 72 66

FAX: 2 239 09 45

Key Personnel, This Location:
Manuel Aguirre Alvardo, General Manager

---

8. CHINA

Beijing Pharmaceutical Factory
Chaoyang District
Beijing 100020

Telephone: 1 502 2077

Hangzhou Minsheng Pharmaceutical Factory
Gongshu, Hangzhou
Zhejiang Province 310011

Telephone: 874 824

Harbin Pharmaceutical Factory
Nangang, Harbin
Heilongdiang Province 150086

Telephone: 63 496

Jining Antiobiotic Factory
Shinzong, Jining
Shandong Province 272121

Telephone: 213 961

Nanfang Pharmaceutical Factory
Luohu District, Shenzen
Guangdong Province 518029

FAX: 360 199

Northeast General Pharmaceutical Works
Tiexi, Shenyang
Liaoning Province 110026

Telephone: 520 133

Shanghai Sino-American Pharmaceutical Co., Ltd.
Minhang District
Shanghai 200024

Telephone: 21 430 2740

Shijiazhuang North China Pharmaceutical Factory
Chang'an Shijiazhuang
Hubei Province 050015

Telephone: 551 133

Southwest Synthetic Pharmaceutical Factory
Jiangbei County, Chongqing
Sichuan Province 631137

Telephone: 811 754 728

Tianjin SmithKline & French Laboratories, Ltd.
Dingjiao
Tianjin 300163

Telephone: 470 622

Wuhan Pharmaceutical Factory
Qiaokou District, Wuhan
Hubei Province 430035

Telephone: 332 281

Xi'an Janssen Pharmaceutical Co., Ltd.
Xincheng, Xi'an
Shanxi Province 710043

Telephone: 334 455

Xiangmin Pharmaceutical Factory
Nantuling
Changsha 400119

Telephone: 731 58 82 11 FAX: 731 42 44 24

Key Personnel, This Location:
Yunde Jiang, Sales Manager

Xinhua Pharmaceutical Factory
Zhangdian, Zibo
Shandong Province 255005

Telephone: 214 223

---

9. COLOMBIA

Biochem Farmacéutica de Colombia, Ltda.
Carrera 41 No. 167-30
Apartado Aéreo 48676
Santéfe de Bogotá, Cundinamarca

Telephone: 1 671 4441

Bristol Farmacéutica, S.A.
Av. 5A Norte No. 26N-80
Apartado Aéreo 249
Cali, Valle

Telephone: 23 67 50 14 FAX: 23 67 14 36

Key Personnel, This Location:
Luis Carlos Quintero, General Manager

Frosst Laboratories, Inc.
Apartado Aéreo 12350-14445
Santafé de Bogotá, Cundinamarca

Telephone: 1 218 46 09 FAX: 1 218 46 16

Key Personnel, This Location:
Alexander Botelho, Manager

Laboartorio Farmédico
Carrera 31 No. 10-190 Arroyo-Hondo
Cali, Valle

Telephone: 23 65 02 87 FAX: 23 65 02 86

Laboratorios Abbott de Colombia
Apartado Aéreo 3589
Santafé de Bogotá, Cundinamarca

Telephone: 1 413 99 00 FAX: 1 413 82 81

Key Personnel, This Location:
Ramiro Zafra-Roldan, General Manager

Laboratorios Biogen de Colombia, S.A.
Carrera 36A No. 24-68
Apartado Aéreoa 82545-100425
Santefé de Bogotá, Cundinamarca

Telephone: 1 269 51 11

Laboratorios Kressfor de Colombia
Calle 17A No. 28A-43
Apartado Aéreo 018491
Santafé de Bogotá, Cundinamarca

Telephone: 1 277 77 62 FAX: 1 237 81 89

Key Personnel, This Location:
Fernando Gutierrez-Cancino, Manager

Laboratorios La Santé, S.A.
Calle 16 No. 32-34
Apartado Aéreo 058696
Santafé de Bogotá, Cundinamarca

Telephone: 1 360 12 60 FAX: 1 237 40 38

Laboratorios Legrand, S.A.
Carrera 20 No. 90-13
Apartado Aéreo 14510
Santafé de Bogotá, Cundinamarca

Telephone: 1 610 49 88 FAX: 1 218 54 26

Laboratorios Stiefel Colombia, S.A.
Carrera 12 No. 98-09
Apartado Aéreo 91492
Santafé de Bogotá, Cundinamarca

Telephone: 1 222 82 73 FAX: 1 618 03 54

Eli Lilly Interamerica, Inc.
Calle 33A Norte No. 3N-13
Apartado Aéreo 4365
Cali, Valle

Parent Company:
Eli Lilly & Company

Telephone: 23 67 15 62 FAX: 23 67 85 53

Key Personnel, This Location:
Victor A Pedroza-Campo, General Manager

Productos Roche, S.A.
Carrera 44 No. 17-21
Apartado Aéreo 14510
Santafé de Bogotá, Cundinamarca

Telephone: 1 244 72 00

SmithKline Beecham
Calle 23 No 7-39
Apartado Aéreo 156-NAL. 287
Cali, Valle

Telephone: 23 82 55 55

Zambon Colombia S.A.
Apartado Aéreo 3145
Santafé de Bogotá, Cundinamarca

Telephone: 1 277 11 37 FAX: 1 201 61 27

Key Personnel, This Location:
Germain Caicedo, Manager

Zyma, S.A.
Calle 11 No. 65-51
Apartado Aéreo 14510
Santafé de Bogotá, Cundinamarca

Parent Company:
Ciba-Geigy AG

Telephone: 1 261 2277 FAX: 1 262 07 78

---

10. DENMARK

Ercopharm A/S
Bogeskovvej 9
Kvistgaard 3490

Telephone: 49 138 342 FAX: 49 138 062

Key Personnel, This Location:
Henning B Termansen, Managing Director

Grindsted Products A/S
Tarnvej 25
Grindsted 7200

Telephone: 50 320 911

Novo Nordisk A/S
Novo Alle
Bagsvaerd 2880

Telephone: 41 448 888

Key Personnel, This Location:
Erik Sorensen, Exec. Vice President
Claus Kuhl, Vice President

Pharmacia A/S
Herredsvejen
Hillerod 3400

Parent Company:
Pharmacia AB

Telephone: 42 265 200

---

11. EGYPT

Amoun Pharmaceutical Industries
125 El Hegaz Street
Heliopolis, Cairo

Telephone: 2 280 6908 FAX: 2 280 6744

Key Personnel, This Location:
Sarwat Bassily, Managing Director

Arab Otsuka Pharmaceutical
66 Cairo
Cairo

Telephone: 2 262 1526

Bristol-Myers Squibb Egypt
17 Studio Misr Street Off Pyramids Street
Giza, Egypt

Telephone: 2 360 6290 FAX: 2 342 2590

Chemipharm, Ltd.
13 Orabi Street
Mohandessin, Giza, Egypt

Telephone: 2 347 0549 FAX: 2 347 4506

Glaxo Abi S.A.E.
26 Sherif Street, Immobilia Building
Cairo

Telephone: 2 280 7100 FAX: 2 280 6031

Merck Sharp & Dohme Egypt
19 Bustan Street, P.O. Box 1029
Cairo

Telephone: 2 393 0233 FAX: 2 392 2891

Pfizer Egypt S.A.E.
17 Ramses Street
Cairo

Telephone: 2 741 474 FAX: 2 767 305

Squibb Egypt
3 Ahmed Nessim Street
Cairo

Telephone: 2 853 716 FAX: 2 536 950

Zeinson Company
19 El Khalifa El Maamoun Street
Cairo

Telephone: 2 668 033 FAX: 2 673 448

---

12. ETHIOPIA

Ethiopian Pharmaceuticals & Medical Supplies Corp.
P.O. Box 21904
Addis Ababa

Telephone: 1 136 059

---

13. FINLAND

Farmos Yhtymae Oy
Tengstoeminkatu 6-8
20101 Turku

Leiras Oy
P.O. Box 415
20101 Turku

Telephone: 21 633 2103 FAX: 21 633 2800

Orion Oy
P.O. Box 425
20101 Turku

Telephone: 21 662 211 FAX: 21 662 545

Key Personnel, This Location:
Matti Kavetvio, Managing Director

Pharmacal Oy
Elimäenkatu 24
00510 Helsinki

Telephone: 0 1464 900 FAX: 0 7530 401

Pharmacia Oy
Box 43 Gran Kulla
Kauniainen, Uusimaa SF-0270

Telephone: 0 505 114

Parent Company:
Pharmacia AB

Tamro Oy
Rajatorpantie 41b, P.O. Box 11
01641 Vantaa

Telephone: 0 852 011

Parent Company:
Spontel Oy

FAX: 0 851 010 10

---

14. FRANCE

Abbott France S.A.
3 Rue de l'Isle
28380 Saint Remy Sur Avre

Parent Company:
Abbott Laboratories

Telephone: 37 48 90 02

FAX: 37 48 91 29

Key Personnel, This Location:
Bertrand Le Gall, Commercial Director

Arkopharma
Zi 1ere Ave 2709M
06516 Carros, Alpes-Maritimes

Telephone: 93 29 11 28

Baxter
Avenue Louis Pasteur
78311 Maurepas Cedex

Telephone: (1) 34 61 50 50

FAX: (1) 34 61 50 25

Key Personnel, This Location:
Danielle Garcia, Director

Bayer Pharma
Rue Bellocier
89103 Sens

Parent Company:
Bayer SA

Telephone: 86 95 71 00

FAX: 86 64 65 05

Key Personnel, This Location:
Franz Schonig, President

Bioron
20 Rue De La Liberation
69110 Ste. Foy Les Lyon, Rhône

Biosedra S.A.
6 Rue du Rempart
27400 Louviers

Telephone: 32 40 36 66

FAX: 32 40 45 03

Key Personnel, This Location:
Pascal Meiller, Director

Bristol-Myers Squibb
Les Collines de l'Arches
92057 Paris, La Defense

Telephone: (1) 47 78 79 30

FAX: (1) 49 00 14 51

Key Personnel, This Location:
Danielle Gerald, Commercial Director

Choay Laboratoire
Rue Du President Salvador Allende
94250 Gentilly, Val-De-Marne

Parent Company:
Choay SA

Telephone: (1) 40 73 40 73

Ciba-Geigy S.A.
37 Ave. De Dale
68330 Huningue, Haut Rhin

Telephone: 89 69 59 69

Parent Company:
Ciba-Geigy AG

Cooperation Pharmaceutique Francaise
Place Lucien Auvert
75016 Ville Dep

Telephone: (1) 64 39 95 50

Delmas Perfusion S.A.
La Baraudiere, B.P. 241
37170 Chambray Les Tours

Telephone: (1) 47 48 43 00 FAX: (1) 47 48 43 14

Key Personnel, This Location:
Ariel Aguettant, President

Elf Sanofi S.A.
10 Avenue Georges V
75008 Paris

Telephone: (1) 40 73 40 73 FAX: (1) 40 73 29 58

Key Personnel, This Location:
Jean F. Dehecq, President

ICI Pharma
1 Rue Des Chauffours
95022 Cergy

Parent Company:
ICI France

Telephone: (1) 34 22 33 33
FAX: (1) 30 38 29 48

Key Personnel, This Location:
Bruno Angelici, President

Industrielle Pharmaceutique S.A.
34 Rue Saint Romain
69008 Lyon

Telephone: 78 77 18 18
FAX: 78 75 39 05

Key Personnel, This Location:
Marcel Grand, Director

Laboratoire Fournier S.A.
Rue des Cortots
21121 Fontaine Les Dijon

Telephone: 80 44 78 00
FAX: 80 44 78 38

Key Personnel, This Location:
Daniel Clausse, President

Laboratoire Fumouze S.A.
7 Place des Martyrs
92113 Clichy

Telephone: (1) 49 68 11 00
FAX: (1) 49 68 41 40

Key Personnel, This Location:
Francois Depoil, President

Laboratoire Guerbet S.A.
16-24 Rue Jean Chaptal
93601 Aulnay-Sous-Bois

Telephone: (1) 45 91 50 00 FAX: (1) 45 91 51 99

Key Personnel, This Location:
Michel Guerbet, President

Laboratoire Pharmacience S.A.
73 Boulevard de la Mission Marchand
92400 Bourbevoir

Telephone: (1) 43 34 31 11 FAX: (1) 43 33 99 37

Key Personnel, This Location:
Claude Guillon, President

Laboratoires Biocodex S.A.
19 Rue Barbes
91126 Montrouge Cedex

Telephone: (1) 46 56 67 89 FAX: (1) 40 92 17 61

Key Personnel, This Location:
Guy Damm, Director

Laboratoires Cassenne
1 Terrasse Bellini
Puteaux, Hauts-De-Seine, 92800

Telephone: (1) 40 81 50 42

Laboratoires Debat
26 Rue Armengaud
St. Cloud, Hauts-De-Seine, 92210

Telephone: (1) 47 11 88 00

Laboratoires Delagrange
Ave. Pierre Brossolette
91380 Chilly Mazarin, Essonne

Telephone: (1) 69 10 77 77

Laboratoires Delalande
16 Rue Henri Regnault
92400 Courbevoie, Hauts-de-Seine

Telephone: (1) 47 89 65 65

Laboratoires Dolisos
71 Rue Beaubourg
75003 Paris, Ville de Paris

Telephone: (1) 42 71 96 16

Laboratoires Fisons
Tour PFA
69130 Hauts-de-Seine

Telephone: (1) 72 19 20 00

Parent Company:
Fisons plc

Laboratoires Glaxo
43 Rue Vineuse
75016 Paris, Ville de Paris

Telephone: (1) 47 55 33 00

Parent Company:
Glaxo Group, plc

Laboratoires Janssen
17 Rue De L'Ancienne Mairie
Boulogne Billancourt,
92100 Hauts-de-Siens

Telephone: (1) 49 09 62 63

Parent Company:
Johnson&Johnson

Laboratoires Riker S.A.
3 Rue Danton
92245 Malakoff Cedex

Telephone: (1) 49 65 51 51
FAX: (1) 49 65 51 00

Key Personnel, This Location:
Pierre Aghassian, President

Laboratoires Roche Posay S.A.
Avenue Rene Lavayer
86270 La Roche Posay

Telephone: 49 86 20 12
FAX: 49 86 14 45

Key Personnel, This Location:
Henri Georges Muller, Director

Laboratoires Sarget
Ave. Du President J. F. Kennedy
33701 Merignac

Telephone: 56 55 35 55 FAX: 56 34 29 24

Key Personnel, This Location:
Christian Senac, President

Laboratoires Upjohn S.A.R.L.
Tour Franklin
92081 Paris La Defense

Parent Company:
The Upjohn Company

Telephone: (1) 47 76 41 44 FAX: (1) 49 00 06 66

Key Personnel, This Location:
Jacques Ory, Manager

Laboratoire Syntex
20 Rue Jean Jaures
92800 Puteaux

Telephone: (1) 47 76 41 23

Key Personnel, This Location:
Dominique Mangeot, President

Laboratoire Unicet
92 Rue Baudin
92307 Lavallois Perret

Telephone: (1) 47 39 94 80 FAX: (1) 47 39 95 69

Key Personnel, This Location:
Jacques Raussou, Secretary General

Les Laboratoires Beecham
6 Esolanade Charles De Gaulle
92000 Nanterre, Hauts-De-Seine

Parent Company:
Beecham Group plc

Telephone: (1) 46 98 46 98

Lilly France S.A.
203 Bureaux De La Colline
92213 St. Cloud, Hauts-de-Seine

Parent Company:
Eli Lilly S.A.

Telephone: (1) 49 11 34 34

FAX: (1) 46 02 27 67

Key Personnel, This Location:
Felix Mosbacher, President

3M Sante
Zone Industrielle
45300 Pithiviers, Loiret

Parent Company:
3M Company

Telephone: (1) 49 65 51 51

Merck Sharp Dohme et Chibret
3 Avenue Hoche
75008 Paris, Ville de Paris

Telephone: (1) 47 54 87 00

FAX: (1) 47 66 99 91

Key Personnel, This Location:
Timothy Fleeth, Director of Marketing

Merrell Dow France S.A.
130 Rue Victor Hugo
92300 Lavallois Perret, Hauts-de-Seine

Parent Company:
Merrell Dow Pharmaceuticals

Telephone: (1) 47 56 33 33 FAX: (1) 47 56 93 93

Key Personnel, This Location:
Pierre Julien, President

Negma Laboratories S.A.R.L.
584 Rue Fourny
78530 Buc

Telephone: (1) 39 56 81 00 FAX: (1) 98 53 77 06

Key Personnel, This Location:
Gabriel Harari, Manager

NP Pharm S.A.
54 Bis Route de Paris
78550 Bazainville

Telephone: (1) 34 87 78 97 FAX: (1) 34 87 78 96

Parke-Davis
11 Ave Dubonnet
92407 Courbevoir, Hauts-de-Seine

Parent Company:
Warner-Lambert Co.

Telephone: (1) 49 01 07 01 FAX: (1) 49 04 07 33

Key Personnel, This Location:
Yves Davin, Director of Research

Pfizer France S.A.
86 Rue de Paris
91407 Orsay

Telephone: (1) 69 85 28 28 FAX: (1) 69 07 75 38

Key Personnel, This Location:
Jean Michel Halfon, Commercial Manager

Pierre Fabre S.A.
45 Place Abel Gance
92654 Bologne

Telephone: (1) 49 10 80 00 FAX: (1) 49 10 97 12

Key Personnel, This Location:
Pierre Fabre, President

Produits Roche
52 Blvd. Du Parc
92521 Neuilly Sur Seine

Parent Company:
Hoffmann LaRoche France S.A.

Telephone: (1) 46 40 50 00 FAX: (1) 47 47 92 95

Key Personnel, This Location:
Claude Schreiner, President

Prographarm
Z.I.St. Arnoult
28170 Châteauneue-En-Thymerais

Telephone: 37 51 86 02 FAX: 37 51 02 88

Promedica S.A.
41 Rue Camille Pelletan
92305 Lavellois Perret

Telephone: (1) 47 37 78 76

Roussel Uclaf S.A.
1 Terasse Bellini
92080 Paris La Defence

Telephone: (1) 40 81 48 87 FAX: (1) 47 78 74 92

Ste. Des Laboratoires Fisch
46 Ave. De La Gare
Vibraye, Sarthe, 72320

Telephone: 43 60 25 25

Schering-Plough
92 Rue Baudin
92307 Levallois Perret, Hauts-de-Seine

Telephone: (1) 47 39 94 80

Septodont S.A.
58 Rue Du Pont de Creteil
94100 St. Maur Des Fosses, Val-de-Marne

Telephone: (1) 48 86 11 77 FAX: (1) 48 85 54 01

Key Personnel, This Location:
Dominique Lablanc, Research Director

Substantia S.A.
11 Avenue Dubonnet
92107 Courbevoie

Telephone: (1) 19 04 07 01

Key Personnel, This Location:
Patrick Lamort de Gail, President

---

15. GERMANY

Abbott GmbH
Max-Planc-Ring 2
65205 Wiesbaden-Delkenheim, Hesse

Parent Company:
Abbott Laboratories

Telephone: (6122) 501 00
FAX: (6122) 501 244

Key Personnel, This Location:
Heinz Kipper, Managing Director

Albert-Roussel Pharma GmbH
Abraham-Lincoln-Strasse 38-42
65189 Wiesbaden

Telephone: (611) 7903-0
FAX: (611) 72 10 89

Key Personnel, This Location:
Ralf Retzlaff, Managing Director

Altana AG
Seedammweg 55
61352 Bad Homberg

Telephone: (6172) 404 0
FAX: (6172) 404 338

Key Personnel, This Location:
H. Bull, CEO

Asche AG
Fischers Allee 49-59
22763 Hamburg, Schleswig-Holstein

Telephone: (40) 398 221 FAX: (40) 398 222 12

Key Personnel, This Location:
H.-J. Klemer

Asta Medica AG
Weissmullerstrasse 45
Postfach 100105
60001Frankfurt am Main, Hesse

Telephone: (69) 400 101 FAX: (69) 127 40

Key Personnel, This Location:
Rolf Kramer, Managing Director

Astra Chemicals GmbH
Tinsdaler Weg 183
22880 Wedel, Schleswig-Holstein

Parent Company:
Astra AB

Telephone: (4103) 708 0 FAX: (4103) 708 293

Key Personnel, This Location:
Andreas Feulner, Managing Director

Azupharma GmbH
Dieselstrasse 5
70839 Gerlingen

Telephone: (7156) 943 0 FAX: (7156) 943 100

Key Personnel, This Location:
K.-P. Reich, Managing Director

Bayer AG
Bayerwerk
50900 Leverkusen, North Rhine-Westphalia

Telephone: (21) 430 1

Bayrol Chemische Fabrik GmbH
Lochhamer Strasse 29
82152 Planegg

Telephone: (89) 857 01 0 FAX: (89) 857 01 777

Key Personnel, This Location:
U. Heyn, Director of Marketing

Behringwerke AG
Postfach 1140
35001 Marburg

Telephone: (6421) 39 0 FAX: (6421) 31 388

Key Personnel, This Location:
Gregor Schulz, Director

Biotest Pharma GmbH
Landsteiner Strasse 3
60720 Dreieich, Hesse

Telephone: (6103) 80 10

Bode Chemie GmbH & Company
Melanchthonstrasse 27
22525 Hamburg, Schelswig-Holstein

Telephone: (40) 540 06 0 FAX: (40) 540 06 200

Key Personnel, This Location:
M. Orlick, Manager

Boehringer Ingelheim KG
Binger Strasse 173, Postfach 200
55218 Ingelheim, Rhineland-Palatinate

Telephone: (6132) 77 0 FAX: (1632) 77 3000

Key Personnel, This Location:
Herbert Johann, Managing Director
Rolf Krebs, Managing Director

Boehringer Mannheim GmbH
Sandhofer Strasse 116
68305 Mannheim, Baden Württemburg

Telephone: (621) 759 0 FAX: (621) 759 2890

Key Personnel, This Location:
Gerhard Musshafen, Director of Sales

Bristol-Myers Squibb GmbH
Volkartstraasse 83
80636 Munich, Bavaria

Parent Company:.
E.R. Squibb & Sons, Inc

Telephone: (89) 121 42 204 FAX: (89) 121 42 442

Key Personnel, This Location:
Joachim von Roy, Chairman

Byk Gulden Lomberg Chemische Fabrik GmbH
Byk-Gulden-Strasse 2
77500 Konstanz, Baden-Württemberg

Telephone: (753) 18 40

Caremark Deutschland GmbH
Zeppelinstrasse 1
85375 Neufahrn

Telephone: (8165) 94 00

Key Personnel, This Location:
H. Stockinger, General Manager

Cascan GmbH & Company KG
Hohenstaufenstrasse 7
65189 Weisbaden, Hesse

Telephone: (611) 793 0 FAX: (611) 793 172

Key Personnel, This Location:
G. Klages, Manager

Cassella-Riedel Pharma GmbH
Hanauer Landstrasse 521
60386 Frankfurt am Main, Hesse

Telephone: (69) 401 20 0 FAX: (69) 401 25 277

Key Personnel, This Location:
P. Schlikker, Marketing and Sales Manager

Cefak Arzneimittel
Ostbahnhofstrasse 15
Postfach 1360
87437 Kempten, Bavaria

Telephone: (831) 74 33 FAX: (831) 77 861

Key Personnel, This Location:
D. Brand, Director

Chephasaar GmbH
Postfach 4120
66376 Sankt Ingbert

Telephone: (6894) 59 09 0 FAX: (6894) 59 09 99

Duphar Pharma GmbH & Company KG
Freundallee 19-23, Postfach 1605
30173 Hannover, Lower Saxony

Telephone: (511) 280 78 0 FAX: (511) 280 78 88

Key Personnel, This Location:
Gerd Mueller, Director of Sales

Durachemie GmbH & Company KG
Pfaffenrieder Strasse 7
82515 Wolfratshausen

Telephone: (8171) 212 03 FAX: (8171) 212 600

Key Personnel, This Location:
E. Roth, Managing Director

Karl Engelhard Fabrik Pharmazeutische Praeparate GmbH & Company KG
Sandweg 94
60316 Frankfurt am Main, Hesse

Telephone: (69) 405 030 50 FAX: (69) 405 030 40

Key Personnel, This Location:
K.-M. Engelhard, Managing Director

Eurim-Pharm Arzneimittle GmbH
Am Gaenselehen 4-5
83451 Piding

Telephone: (8651) 704 0 FAX: (8651) 704 11

Key Personnel, This Location:
H. Lipp, Sales Manager

Fresenius AG
Borkenberg 14
61440 Oberursel, Hesse

Telephone: (6171) 60 0 FAX: (6171) 266 39

Key Personnel, This Location:
A. Paech, Sales Manager

Glaxo GmbH
Alsterufer 1
20354 Hamburg, Schleswig-Holstein

Telephone: (40) 415 23 0 FAX: (40) 448 07 15

Key Personnel, This Location:
H. Davis, Managing Director

Gruenenthal GmbH
Steinfeldstrasse 2
52222 Stolberg, North Rhine-Westphalia

Telephone: (2402) 103 1 FAX: (2402) 103 449

Key Personnel, This Location:
Dr. Egly, Managing Director

Heyl Chemisch Pharmazeutische Fabrik Gmbh & Company KG
Goerzalle 253
14167 Berlin, Berlin

Telephone: (30) 816 96 0 FAX: (30) 817 40 49

Key Personnel, This Location:
E. Heyl, Managing Director

Hexal Pharma GmbH & Company KG
Postfach 1269
83607 Holzkirchen

Telephone: (8024) 30 05 0 FAX: (8024) 30 05 55

Key Personnel, This Location:
H. Hardacker, Managing Director

Hoechst AG
Postfach 80 03 20
65926 Frankfurt am Main, Hesse

Telephone: (69) 305 0 FAX: (69) 357 177

Key Personnel, This Location:
Wolfgang Hilger, Chairman
Guenter Metz, Deputy Chairman

Hoffmann-LaRoche AG
Emil-Barell-Strasse 1
79639 Grenzach-Wyhlen, Baden-Württemberg

Telephone: (7624) 14 0 FAX: (7624) 10 19

Key Personnel, This Location:
R. Maertens, Director of Sales

Ichthyol Gesellschaft
Sportallee 85
22335 Hamburg, Schleswig-Holstein

Telephone: (40) 507 14 0 FAX: (40) 507 14 110

Key Personnel, This Location:
D. Filbry, Director

ICI Pharma Arzneimittelwerk
Otto-Hahn Strasse
68723 Plankstadt

Telephone: (6202) 82 00 FAX: (6202) 10 977

Kessla Chemie GmbH, **Pharma Wolfen**
Greppinger Strasse 20a
06766 Wolfen

Telephone: (3493) 778 53 FAX: (3493) 776 10

Key Personnel, This Location:
George W. Weber, Director of Marketing

Knoll AG
Postfach 210805
67008 Ludwigshafen, Rhineland-Palatinate

Telephone: (621) 589 0 FAX: (621) 589 28 96

Lilly Deutschland GmbH
Saalburgstrasse 153
61350 Bad Homburg, Hesse

Telephone: (6172) 273 0 FAX: (6172) 273 283

Key Personnel, This Location:
D. Welzel, Managing Director

Lipha Arzneimittel GmbH
Zeche Katarina 6, Postfach 130326
45307 Essen, North Rhine-Westphalia

Telephone: (201) 81 00 20 0 FAX: (201) 81 00 21 3

Madaus AG
Ostmerheimer Strasse 198
51109 Cologne, North Rhine-Westphalia

Telephone: (221) 89 98 1 FAX: (221) 89 98 305

Key Personnel, This Location:
U. Madaus, Vice President

Marion Merrell Dow GmbH
Eisenstrasse 40
65428 Russelsheim

Telephone: (6142) 691 0 FAX: (6142) 691 105

E. Merck Chemische Fabrik GmbH
Frankfurter Strasse 250
64293 Darmstadt, Hesse

Telephone: (6151) 72 0 FAX: (6151) 72 20 00

Key Personnel, This Location:
M. May, Director Sales & Marketing

Nordmark Arzneimittel GmbH
Pinnauallee 4, Postfach 1224
25430 Uetersen, Schleswig-Holstein

Parent Company:
Knoll AG

Telephone: (4122) 712 0 FAX: (4122) 712 220

Key Personnel, This Location:
C.H. Pich, Managing Director

Nycomed Arzneimittel GmbH
Freisinger Landstrasse 74
80939 Munich, Bavaria

Telephone: (89) 323 98 0 FAX: (89) 323 98 444

Key Personnel, This Location:
H. Rentzsch, Director Sales

K.W. Pfannenschmidt GmbH
Offakamp 9
22529 Hamburg, Schleswig-Holstein

Telephone: (40) 553 10 57 FAX: (40) 553 64 14

Key Personnel, This Location:
H. Mundt, General Manager

Pfizer GmbH
Pfizerstrasse 1
76139 Karlsruhe, Baden-Württemberg

Telephone: (721) 610 10 1 FAX: (721) 610 14 36

Key Personnel, This Location:
Werner Soukup, Director of Pharmaceuticals

Pfleger GmbH
Dr. Robert-Pfleger Strasse 12
96052 Bamberg, Bavaria

Telephone: (951) 604 30 FAX: (951) 604 329

Key Personnel, This Location:
H. Pfleger, Managing Director

Sandoz AG
Deutschherrnstrasse 15
Postfach 8500
90429 Nürnberg, Bavaria

Telephone: (911) 273 0 FAX: (911) 273 653

Schering AG
Mullerstrasse 178
1623 Berlin, Berlin

Telephone: (30) 468 0 FAX: (30) 468 53 05

Key Personnel, This Location:
Guisseppe Vita, Chairman

Siegfried Pharma
Mumpferfaehrstrasse 68
79713 Bad Saeckingen

Telephone: (7761) 559 0 FAX: (7761) 854 1

Key Personnel, This Location:
H. Kunz, Manager

SmithKline Beecham Pharma GmbH
Sapporobogen 6-8
80809 Munich, Bavaria

Telephone: (89) 157 05 0 FAX: (89) 157 05 123

Key Personnel, This Location:
H.J. Bruehr, Director of Sales & Marketing

Stada AG
Stadastrasse 2-18, Postfach 1260
61118 Bad Vilbel

Telephone: (6101) 603 0 FAX: (6101) 603 259

Key Personnel, This Location:
R. Legien, Manager

Synthapharm GmbH
Delle 50-52
45468 Muehlheim-Ruhr

Telephone: (208) 350 94 FAX: (208) 382 471

Key Personnel, This Location:
H. Lier, Sales Manager

TAD Pharmazeutisches Werk GmbH
Heinz-Lohmann-Strasse 5
27472 Cuxhaven, Lower Saxony

Telephone: (4721) 606 0 FAX: (4721) 606 333

Key Personnel, This Location:
H. Floettmann, Director

Dr. Karl Thomae GmbH
Birkendorfer Strasse 65
88400 Biberach on der Riss, Baden-Württemberg

Telephone: (7351) 54 0 FAX: (7351) 54 46 00

Key Personnel, This Location:
H. Buerk, Manager

Upjohn GmbH
Humboltstrasse 10
64646 Heppenheim

Telephone: (6252) 704 0 FAX: (6252) 704 291

Key Personnel, This Location:
F.-J. Brunnenberg, Development Director

Wellcome GmbH
Im Langen Felde 3-5
30938 Burgwedel

Parent Company:
The Wellcome Foundation, Ltd.

Telephone: (5139) 804 0
FAX: (5139) 804 269

Key Personnel, This Location:
F. Frauendorfer, Director of Sales & Marketing

Wolff GmbH & Company KG
Sudbrackstrasse 56
33611 Bielefeld, North Rhine-Westphalia

Telephone: (521) 880 80 5
FAX: (521) 880 83 34

Key Personnel, This Location:
Michael Harras, Director of Sales

Wyeth Pharma GmbH
Schleebruggenkamp 15
48159 Munster, North Rhine-Westphalia

Telephone: (251) 204 0
FAX: (251) 204 128

Key Personnel, This Location:
L. Gastmeyer, Marketing Manager

---

16. GREECE

Adelko Kolokotronis E. Bros.
37 Pireos Street
Athens GR 183 46

Telephone: 1 481 9311 FAX: 1 481 5790

Key Personnel, This Location:
E. Kolokotronis, President

Cana Pharmaceutical Laboratory S.A.
446 Herakleiou Avenue
Athens GR 141 22

Telephone: 1 281 5502 FAX: 1 283 0773

Key Personnel, This Location:
Vassili Niadas, Director

Famar S.A.
63 Aghiou Demetriou Street
Athens GR 115 21

Telephone: 1 982 5521 FAX: 1 981 0908

Key Personnel, This Location:
J. Marinopoulos, Managing Director

Farmalex S.A.
15-17 Tsocha
Athens GR 115 10

Telephone: 1 644 5612 FAX: 1 644 5690

Key Personnel, This Location:
V. Alexakos, Director

P. N. Gerolymatos S.A.
35 Mihalakopoulou Street
Attica GR 115 28

Telephone: 1 724 8193 FAX: 1 723 8976

Key Personnel, This Location:
Panoyotis N. Gerolymatos, Managing Director

Glaxo S.A.
226 Ionias Avenue
Acharnon, Athens 111 44

Hoechst Hellas AG
Tatoiou Rd
Nea Erythraia 146 71

Telephone: 1 202 5513

B. Niadas & Sons S.A.
16 Kifissou Avenue
Athens GR 182 33

Telephone: 1 481 9106 FAX: 1 481 0021

Key Personnel, This Location:
V. J. Niadas, General Manager

Pharmaserve S.A.C.I.
335 Messogion Avenue
Athens GR 153 10

Telephone: 1 572 6391 FAX: 1 647 7771

Key Personnel, This Location:
Dionysios Filiotis, Managing Director

Vianex S.A.
19 Satovriandou Street
Athens GR 104 31

---

17. GUATEMALA

Abbott Laboratories
Calzada Roosevelt 6-19
Colonia Cotio
Zona 3, Mixco

Parent Company:
Abbott Laboratories

Telephone: 9 10 11 15

FAX: 9 32 24 19

Key Personnel, This Location:
Dario Castillo, Director

Laboratorios Quimicos Farmaceuticos Lancasco S.A.
Km 55 Ruta Atlantico Zona 18
Guatemala City, Guatamala

Laprin S.A.
Apartado Postal 2201
Guatamala City, Guatamala 01062

Telephone: 2 934 1921

FAX: 2 934 1929

Key Personnel, This Location:
Geraldina F. de Samayoa, Manager

Quimica Hoechst De Guatemala S.A.
Carretera Roosvelt Km 15 5 Zona
Guatemala City, Guatamala

---

18. HONG KONG

C. Amy Laboratories, Ltd.
Units 1-4 First Floor, Eastern Centre
1065 King's Road

Telephone: 562 1289 FAX: 565 7913

Key Personnel, This Location:
Lau Ho Kung, Marketing Manager

Colin McKenzie & Co., Ltd.
701 Hammer Hill Road
Kowloon

Telephone: 322 0719 FAX: 329 4527

Key Personnel, This Location:
Zung Gee Liang, Managing Director

Connor Pharmaceuticals, Ltd.
12/F, Block B, E On Factory Building
14 Wong Chuk Hang Road
Aberdeen

Telephone: 553 4229 FAX: 814 0197

Key Personnel, This Location:
Jeffrey Sun, Director

Jean Marie Pharmacal Company, Ltd.
78-84 Wang Lung Street, 17th Floor
Jing Ho Industrial Building
Tsuen Wan, N.T.

Telephone: 407 3271 FAX: 407 5707

Key Personnel, This Location:
Strauss Chu, Director

Loyal Advance, Ltd.
8/F, B2 & 4, Phase 1
Tsing Yi Industrial Centre, N.T.

Telephone: 433 6608 FAX: 433 6777

Key Personnel, This Location:
Chan Chung Fu, Production Manager

Marching Pharmaceutical, Ltd.
9/F, Room 9-12 Vigor Industrial Building, Blk. B
14-20 Cheung Tat Road
Tsing Yi Island, N.T.

Telephone: 432 6993 FAX: 432 3522

Key Personnel, This Location:
Yu Lam, Director

---

19. HUNGARY

Abbott
Delej ytca 20
1089 Budapest

Parent Company:
Abbott Laboratories, Inc.

Telephone: (1) 267 1014
FAX: (1) 267 1013

Key Personnel, This Location:
Todd Savidge, Managing Director

Alkaloida Chemical Company, Ltd.
Kabay J u. 29
4440 Tiszavasvâri

Telephone: (42) 372 511
FAX: (42) 372 512

Key Personnel, This Location:
Lazlo Dupcsak, General Manager

Baxter Hungary
Dozsa Gyorgy ût 44
1071 Budapest

Parent Company:
Baxter International, Inc.

Telephone: (1) 142 8993
FAX: (1) 142 9792

Key Personnel, This Location:
Krisztina Czak, Managing Director

Biogal Pharamceutical Works
Pallagi ût 13
4042 Debrecen

Telephone: (52) 311 633 FAX: (52) 318 752

Key Personnel, This Location:
Sandor Arany, President & CEO

Chinoin Pharmaceutical & Chemical Works
To ûtca 1-5
1045 Budapest

Telephone: (1) 169 0900 FAX: (1) 169 0293

Key Personnel, This Location:
Miklos Gyorgy, Director General
Lazlo Kadar, Commercial Director

Egis Pharmaceuticals
Kereszturi ût 30-38
1106 Budapest

Telephone: (1) 252 7222 FAX: (1) 183 4554

Key Personnel, This Location:
Istvan Orban, Director General

Human Vaccine & Pharmaceutical Production
Tancsics Mihaly ût 82
2101 Godollo

Telephone: (28) 320 733 FAX: (28) 320 177

Key Personnel, This Location:
Lajos Aradi, Managing Director

Janssen Pharmaceuticals Hungary
Laborc ût 1
1037 Budapest

Parent Company:
Johnson & Johnson

Telephone: (1) 250 2944
FAX: (1) 188 6373

Key Personnel, This Location:
Francisco Valez, General Manager

Lilly Hungaria Kft
Madach Imre u 13
1075 Budapest

Parent Company:
Eli Lilly & Company

Telephone: (1) 267 9801
FAX: (1) 268 1500

Key Personnel, This Location:
Roland Powell, Managing Director

Phylaxia-Sanofi Vaccine Company
Szallas u. 5
1107 Budapest

Telephone: (1) 157 5311
FAX: (1) 127 4617

Key Personnel, This Location:
Ferenc Czelleng, Director General

Reanal Fine Chemical Works Rt
Telepes u. 53
1147 Budapest

Telephone: (1) 252 1357
FAX: (1) 252 1339

Key Personnel, This Location:
Lazlo Zsigmond, Director General

Gedeon Richter Chemical Works
Gyomroi ût 19-21
1103 Budapest

Telephone: (1) 261 2199 FAX: (1) 260 6650

Key Personnel, This Location:
Erik Bogsch, Director General

---

20. INDIA

Biddle Sawyer, Ltd.
25 Delal Street
Fort, Bombay 400 023

Telephone: 22 273 3686 FAX: 22 273 3263

Camlin, Ltd.
Camlin House, J.B. Nagar Andheri (E)
Bombay 400 059

Telephone: 22 836 6635 FAX: 22 836 6579

Colinz Laboratories, Ltd.
Copper Roller Compound, 2nd Floor
Lbs Marg, Bhandup, Bombay 400 078

Telephone: 22 560 4569 FAX: 22 564 2714

Key Personnel, This Location:
L.S. Mani, Managing Director

Farmson Pharmaceutical Gujarat, Ltd.
Commerce Center, 5th Floor
Sayaji Gunj, Vadodara 390 005

Telephone: 265 330 606 FAX: 265 330 103

IPCA Laboratories, Ltd.
63-E, IPCA House
Kandivil Industrial Estate
Kandivil, Bombay 400 067

Telephone: 22 808 60 97 FAX: 22 808 16 13

J.K. Pharmaceuticals
53 Community Centre
New Friends Colony, Mathura Rd.
New Delhi 110 065

Parent Company:
J.K. Industries, Ltd.

Telephone: 11 682 12 48 FAX: 11 682 12 89

Kopran, Ltd.
Parijat House
1076 Dr. E. Moses Road
Worli, Bombay 400 018

Telephone: 22 492 24 55 FAX: 22 495 03 63

Lyka Laboratories, Ltd.
77 Nehru Road
Bombay 400 099

Telephone: 22 610 59 00 FAX: 22 611 10 24

Micro Labs, Ltd.
303 111 Floor A Wing
Queens Corner Apartments, 3
Queens Road
Bangalore 560 001

Telephone: 80 226 96 09 FAX: 80 226 63 23

The Pharmaceutical Products Of India, Ltd.
212 Unique Industrial Estate
Cardinal Gracious Road
Chakala, Andheri, Bombay 400 099

Telephone: 22 834 44 66 FAX: 22 837 07 72

Pradeep Drug Company, Ltd.
98 Lloyds Road
Madras 600 014

Telephone: 44 827 62 61 FAX: 44 825 42 93

Ranbaxy Laboratories, Ltd.
A-1 Phase 1, Okhla Industrial Area
New Delhi, 110 020

Telephone: 11 683 35 58 FAX: 11 683 53 90

Key Personnel, This Location:
Romi Barat

Themis Chemicals, Ltd.
11/12 Udyognager Industrial Estate
Coregaon (West), Bombay 400 063

Telephone: 22 875 56 76 FAX: 22 874 66 21

Torrent Pharmaceuticals, Ltd.
Torrent House, off Ashram Road
Ahmedabad 380 009

Telephone: 79 405 090 FAX: 79 460 048

Unichem Laboratories, Ltd.
Mahalaxmi Chambers, 22-B Desai Road
Bombay 400 026

Telephone: 22 492 66 14 FAX: 22 495 04 50

USV, Ltd.
Poonam Chambers, b, Dr. A.B. Road
Worli, Bombay 400 018

Telephone: 22 492 00 23 FAX: 22 495 03 81

---

21. INDONESIA

Bristol-Myers Indonesia PT
Menara Duta Bldg.
Th Hr Rasuna Said Kav B9 Kuninggan
Jakarta, Jakarta

ICI Farmasi Indonesia PT
S. Widjojo Center
Jl Jend Sudirman 71
Jakarta, Jakarta 10001

Parent Company:
ICI, Ltd.

Konimex, PT
P.O. Box 33
Solo, 57101

Telephone: 21 333 575 FAX: 21 321 856

Key Personnel, This Location:
Djoenaedi Joesoef, President

Schering-Plough Indonesia PT
Bank Pacific Bldg., 9th Floor
Jakarta, Jakarta 10220

Telephone: 21 570 6912 FAX: 21 570 6791

Key Personnel, This Location:
P.A. Townsend, President

Squibb Indonesia PT
Jalan Cimandiri 24
Jakarta, Jakarta 10001

---

22. IRELAND

Allergan Pharmaceuticals (Ireland), Ltd.
Westport
County Mayo

Telephone: 98 25 222 FAX: 98 25 791

Key Personnel, This Location:
Maureen Cameron

Antigen Pharmaceuticals, Ltd.
Roscrea
County Tipperary

Telephone: 505 21 112

Key Personnel, This Location:
Liam Doorley

Elan Corp., plc
Mooksland, Athlone
County Westmeath

Telephone: 902 94 666 FAX: 902 94 931

Key Personnel, This Location:
D.J. Geaney, President
K. McVey, Exec. Vice President

Leo Laboratories
Cashel Road at Crumlin Road
Dublin, County Dublin

Telephone: 1 490 8924

Key Personnel, This Location:
Eamon O'Neil

Eli Lilly, Ltd.
Dunderrow, Kinsale
County Cork

Telephone: 21 772 699 FAX: 21 775 503

Key Personnel, This Location:
Brenden Scannel

Merck Sharp & Dohme (Ireland), Ltd.
Ballydine, Kilsheelan, Clonmel
County Tipperary

Telephone: 51 404 11 FAX: 51 408 36

Norbrook Laboratories, Ltd.
Station Works, Carnlough Road
Newry

Telephone: 1693 644 35

Key Personnel, This Location:
R. Campbell

Pfizer Pharmaceuticals Production Corp.
Ringaskiddy
County Cork

Telephone: 21 378 701 FAX: 21 378 087

Schering-Plough Co.
Rathdrum
County Wicklow

Telephone: 404 462 09 FAX: 404 462 26

Key Personnel, This Location:
Conor O'Brien

Stafford Miller Ireland, Ltd.
Younghal Road, Dungarven
County Waterford

Telephone: 58 424 22

Key Personnel, This Location:
John Butler

---

23. ISRAEL

Abic, Ltd.
Kiryat Nordau Industrial Zone
Netanya

Telephone: 5 363 97 77

Agis Industries
26 Lehi Street
Bnei Brak

Telephone: 3 577 37 37

Chemagis, Ltd.
3 Hashlosha Street
P.O. Box 9091
Tel-Aviv 61090

Parent Company:
Agis Group

Telephone: 3 537 5121

FAX: 3 537 3090

Taro Pharmaceutical Industries, Ltd.
Bet Merkazim, Maskit Street
Herzila Pituach, 46120

Telephone: 9 557 942

FAX: 9 557 443

Teva Pharmaceuticals
5 Basel Street, P.O. Box 3190
Petach TIkva 49131

Telephone: 3 926 7267 FAX: 3 923 4050

<u>Key Personnel, This Location:</u>
Eli Hurvitz, President
Ben-Ami Rosenfeld, Exec. Vice President

---

24. ITALY

Abbott S.p.A
V. Pontina, Km 22
Campoverde, 04010 (LT)

Parent Company:
Abbott Laboratories

Telephone: 6 92 89 21

FAX: 6 92 53 193

Key Personnel, This Location:
Gino Faedda, President

Allergan S.p.A.
V. Costarica 20/22
Pomezia 00040 (RM)

Telephone: 6 91 18 31

FAX: 6 91 22 270

Key Personnel, This Location:
M.J. Donohoe, President

Archimica S.p.A.
vle. Europa 11
Origgio VA 21040

Telephone: 2 96 36 71

FAX: 2 967 30 45 6

Astifarma S.p.A.
Asti 141 00
Torin 114

Telephone: 141 21 02 82

FAX: 141 21 66 67

Key Personnel, This Location:
Ada Formica, President

Bieffe Medital S.p.A.
v. Nuova Provinciale
Grosotto, 23034 (SO)

Telephone: 342 88 71 91 FAX: 342 88 71 72

Key Personnel, This Location:
Francesco Siccardi, Administrator

Bioresearch S.p.A.
v. Fosse Ardeatine 2
Liscate, 20060 (MI)

Parent Company:
Knoll AG

Telephone: 2 95 87 87 4 FAX: 2 95 87 77 1

Key Personnel, This Location:
Livo Camozzi, President

Boehringer Mannheim Italia S.p.A
v. Sant'Uguzzone 5
Milan 20126

Telephone: 2 25 28 1 FAX: 2 25 28 22 0

Key Personnel, This Location:
Massimilliano Pancera, President

Bristol Myers Squibb S.p.A.
v. Paolo di Dono 73
Rome 00143

Telephone: 6 50 39 61

Key Personnel, This Location:
Guido Porporati, President

Camillo Corvi S.p.A
v. dei Mille 3
Piacenza 29100

Telephone: 523 54 51 FAX: 523 54 52 76

Key Personnel, This Location:
Fluvio Domanico, President

Ciba Geigy S.p.A
Str. Stat. 233, Varesina Km 25,000
Origgio, 21040 (VA)

Telephone: 2 96 54 1 FAX: 2 96 54 29 10

Key Personnel, This Location:
Sergio Giuliani, President
P.A. Borney, Vice President

Comifar S.p.A
v. Mechiorre Gioia 124
Milan 20125

Telephone: 2 68 31 41

Key Personnel, This Location:
Giancarlo Sabbia, Director General

Crinos-Industria Farmacobiologica S.p.A.
p. XX Settembre 2
Villa Guardia, 22079 (CO)

Telephone: 31 48 01 91 FAX: 31 48 17 84

Ellem Industria Farmaceutica s.r.l.
c. di Porta Ticinese 89
Milan 20123

Telephone: 2 58 19 71 FAX: 2 58 19 73 03

Key Personnel, This Location:
Giampaolo Zimbeletti, President

Fidia S.p.A.
v. Ponte della Fabbrica 3/A
Abano Terme Bagni, 35031 (PD)

Telephone: 49 82 32 11 1 FAX: 49 81 03 40

Key Personnel, This Location:
Ennio Arengi, President
Alberto Locatelli, Director General

Gruppo Lepetit S.p.A.
v. del Ghisallo 20
Milan 20151

Telephone: 2 30 02 1FAX: 2 30 02 54 00

Key Personnel, This Location:
Emillio Lepetit, President

Hoechst Italia S.p.A.
v. M.U. Traiano 18
Milan 20149

Telephone: 2 31 07 1FAX: 2 33 11 35 1

Key Personnel, This Location:
Francesco Cimmarrusti, President

ICI-Pharma S.p.A.
Vle. Isonzo 25
Milan 20135

Parent Company:
ICI Italia, S.p.A.

Telephone: 2 58 39 21

FAX: 2 58 30 22 20

Industria Farmaceutica Serono S.p.A.
Via Casilina 125
Rome 00176

Telephone: 6 70 38 41

FAX: 6 70 38 43 43

Key Personnel, This Location:
Giovanni Battista Arduino, President

Inverni Della Beffa
Via Ripamonti 99
Milan 20141

Telephone: 2 57 49 61

FAX: 2 57 40 46 20

Key Personnel, This Location:
Luigi Della Beffa, President

Istituto Gentili S.p.A.
v. Mazzini 112-C/P 328
Pisa 56125

Telephone: 50 92 61 11

FAX: 50 92 63 31

Key Personnel, This Location:
M.G. Gentili Mian,President
Piero P. Neggiani, Technical Director

Janssen Farmaceutici S.p.A.
v. Castello della Magliana
Rome 00148

Telephone: 6 65 58 81 FAX: 6 65 51 17 48

Key Personnel, This Location:
Giovani Luchetti, Director General

Eli Lilly Italia S.p.A.
Via Gramsci 721/733
Sesto Fiorentino, 50019 (FL)

Parent Company:
Eli Lilly S.A.

Telephone: 55 42 57 1 FAX: 55 45 57 701

Key Personnel, This Location:
Robert Anthony Breyer, President

L. I. R. C.A.-Synthelabo s.r.l.
v. Rivoitana 35
Limito, 20090 (MI)

Telephone: 2 92 79 1FAX: 2 92 69 98 31

Key Personnel, This Location:
Francesco Lapeyre, President

Malesci SPA
v. Nicola Porpora 22/24
Florence 50144

Telephone: 55 33 90 0 FAX: 55 36 10 37

Key Personnel, This Location:
Giovanni Soro, President

Menarini Industrie Farmaceutiche Riunite-s.r.l.
v. Sette Santi 3
Florence 50131

Telephone: 55 56 80 1

Key Personnel, This Location:
Alberto Aleotti, Administrator

MSD-Merck Sharp & Dohme (Italia) S.p.A.
v. Giovanni Fabbroni 6
Rome 00191

Telephone: 6 36 19 11 FAX: 6 32 78 61 9

Key Personnel, This Location:
Umberto Mortari, President

Neopharmed S.p.A.
via Pordoi 18- Baranzate
Bollate 20021(MI)

Telephone: 2 33 33 61 FAX: 2 38 20 06 74

Key Personnel, This Location:
Massimo Sordi, President

Parke-Davis S.p.A.
v. C. Colombo 1
Liante, 20020 (MI)

Parent Company:
Warner-Lambert Company

Telephone: 2 93 33 71 FAX: 2 93 33 72 47

Key Personnel, This Location:
Francesco Canstantini, President

Pfizer Italiana S.p.A.
v. Valbondione 113
Rome 00188

Parent Company:
Pfizer Intl. Inc.

Telephone: 6 33 18 2
FAX: 6 30 02 01 9

Key Personnel, This Location:
August Humpert, President

Pierrel S.p.A.
v. Bisceglie 96
Milan 20152

Telephone: 2 41 40 1
FAX: 2 41 40 40 0

Key Personnel, This Location:
Haken Astrom, President

Polifarma S.p.A.
v. Tor Sapienza 138
Rome 00155

Telephone: 6 22 88 74 1
FAX: 6 22 83 12 0

Key Personnel, This Location:
Carlo Materazzi, Technical Director

Procter & Gamble Italia S.p.A.
v. C. Pavese 385
Rome 00144

Parent Company:
Procter & Gamble Company

Telephone: 6 50 09 01
FAX: 6 50 11 88 1

Key Personnel, This Location:
E.J. Pepper, President
Vito Varvaro, Vice President

Recordati Industria Chimica E Farmaceutica S.p.A.
v. M. Civitali 1
Milan 20148

Telephone: 2 48 78 71 FAX: 2 48 70 23 22

Key Personnel, This Location:
Arrigo Recordati, President

Rhône-Poulenc Rorer S.p.A
vle. Europa 11
Origgio, 21040 (VA)

.Parent Company:
Rhône-Poulenc Rorer, Inc.

Telephone: 2 96 10 1 FAX: 2 96 73 05 35

Key Personnel, This Location:
Fabio Enrico Guatelli, President

Roche S.p.A.
Piazza Durante 11
Milan 20131

Telephone: 2 28 84 1 FAX: 2 28 20 60 3

Key Personnel, This Location:
Andres P. Levenberger, President

Sandoz Prodotti Farmaceutici S.p.A.
v. C. Arconati 1
Milan 20135

Parent Company:
Sandoz AG

Telephone: 2 57 95 1 FAX: 2 57 95 30 5

Key Personnel, This Location:
Guiseppe Miglio, Director General

Sanofi Winthrop S.p.A.
v. G.B. Piransei 38
Milan 20137

Telephone: 2 73 94 1 FAX: 2 76 11 03 89

Key Personnel, This Location:
Wolfgang Horedt, President

S.I.F.R.A. - Societa Italiana Farmaceutici Ravizza S.p.A.
v. Camagre 41
Isola Della Scala, 37063 (VR)

Telephone: 45 73 00 61 1 FAX: 45 73 01 10 1

Key Personnel, This Location:
Claudio Cavazza, President

SmithKline Beecham Farmaceutici S.p.A.
v. Zambelletti-Baranzate
Bollate, 20021 (MI)

Parent Company:
SmithKline Beecham Corp.

Telephone: 2 38 06 1 FAX: 2 35 01 88 2

Key Personnel, This Location:
Ambrogio Secondi, Presidente

Sorin Biomedica S.p.A.
Via Crescentino
Saluggia, 13040 (VC)

Telephone: 161 48 71 FAX: 161 48 75 45

Key Personnel, This Location:
Umberto Rosa, President

S.P.A. - Societa Prodotti Antibiotici, S.p.A.
v. Biella 8
Milan 20143

Telephone: 2 81 83 1 FAX: 2 81 32 98 3

Key Personnel, This Location:
Hans Hugo Van den Heuvel, President
Giorgio Pisani, Director General

Sterling Midy S.p.A.
v. Guiseppe Colombo 40
Milan 20133

Telephone: 2 74 83 1 FAX: 2 74 83 22 2

Key Personnel, This Location:
Daniel Duprè, President
Serafino Paita, Director General

Upjohn S.p.A.
v. G.E. Upjohn 2
Caponago, 20040 (MI)

Parent Company:
The Upjohn Co.

Telephone: 2 95 05 21 FAX: 2 95 74 16 66

Key Personnel, This Location:
Luigi Fracas, Director General

Wellcome Italia S.p.A.
v. del Mare 36
Promezia 00040 (RM)

Parent Company:
The Wellcome Foundation, Ltd.

Telephone: 6 91 18 01

Key Personnel, This Location:
Feriano Banci, President

Wyeth S.p.A.
v. Nettunense 90
Aprilla, 04011 (LT)

Parent Company:
American Home Products Corp.

Telephone: 6 92 71 52

FAX: 6 92 39 61

Key Personnel, This Location:
Carlo Tassini, Director General

Zambon Group SPA
v. Lillo Del Duca 10
Bresso, 20091(MI)

Telephone: 2 66 52 41

FAX: 2 66 50 14 92

Key Personnel, This Location:
Mario Fertonani, President

Zeneca S.p.A.
v. Francesco Sfoza Pal. Volta-Milano 3
Basiglio, 20080 (MI)

Telephone: 2 90 45 41

FAX: 2 90 75 56 15

Key Personnel, This Location:
Gianni Marini, President

---

25. JAPAN

Asahi Chemical Industry Company, Ltd.
4-5-13 Shibaura Minato-ku
Tokyo 108

Telephone: 3-3456-8225 FAX: 3-3456-8271

Banyu Pharmaceutical Company, Ltd.
2-3-3 Nihonbashi-Honcho
Chuo-ku Tokyo 103

Telephone: 3-3270-7551 FAX: 3-3246-2130

Bayer Yakuhin, Ltd.
1-8-12 Hon-Machi
Nihon Seimei Bldg.
Chuo-ku Osaka 541

Parent Company:
Bayer AG

Telephone: 6-263-9111 FAX: 6-266-1335

Bristol-Myers Squibb K.K.
2-12-19 Shibuya, Shibuya-ku
Tokyo 150

Telephone: 3-3409-6684 FAX: 3-3486-0605

The Chemo-Sero Threapeutic Research Institute
Okubo, Shimizu-machi
Kumamoto 860

Telephone: 96-344-1211 FAX: 96-345-1345

Chugai Pharmaceutical COmpany, Ltd.
2-1-9 Kyobashi, Chuo-Ku
Tokyo 104

Telephone: 3-3281-6611 FAX: 3-3281-2828

Ciba Geigy (Japan), Ltd.
10-66 Miyuki cho
Takarazuka 0665

Parent Company:
Ciba-Geigy AG

Telephone: 797-74-2381 FAX: 797-74-2544

Daiichi Pharmaceutical Co., Ltd.
3-14-10 Nihonbashi, Chuo-ku
Tokyo 103

Telephone: 3-3272-0611 FAX: 3-3281-8427

Dainippon Pharmaceutical Company, Ltd.
2-6-8 Dosho-machi, Chuo-ku
Osaka 541

Telephone: 6-203-5321 FAX: 6-203-6581

Eiken Chemical Company, Ltd.
1-33-8 Hongo, Bunkyo-Ku
Tokyo 113

Telephone: 3-3818-5401 FAX: 3-3818-1207

Eisai Company, Ltd.
4-6-10 Koishikawa, Bunkyo-ku
Tokyo 112-88

Telephone: 3-3817-3700 FAX: 3-3811-7420

Fujimoto Pharmaceutical Corp.
1-3-40 Nishiotsuka
Matsubara City, Osaka 580

Telephone: 723-32-5151 FAX: 723-34-6757

Fujirebio, Inc.
2-7-1 Nishi-Shinjuku, Shinjuku-ku
Tokyo 163

Telephone: 3-3348-0691 FAX: 3-3342-6220

Fujisawa Astra, Ltd.
3-6-8 Kyutaro-Machi, Chuo-ku
Osaka 541

Telephone: 6-244-7810 FAX: 6-244-7860

Fujisawa Pharmaceutical Co., Ltd.
3-4-7 Dosho-macho, Chuo-ku
Osaka 541

Telephone: 6-202-1141 FAX: 6-222-4988

Fuso Pharmaceutical Industries, Ltd.
2-3-11 Morinomiya, Joto-ku
Osaka 536

Telephone: 6-969-1131 FAX: 6-961-0061

The Green Cross Corp.
1-3-3 Imabashi, Chuo-ku
Osaka 541

Telephone: 6-227-4625 FAX: 6-227-0782

Grelan Pharmaceutical Company, Ltd
2-12-3 Sakurashin-machi, Setagaya-ku
Tokyo 154

Telephone: 3-3429-5281 FAX: 3-3429-6997

Hisamitsu Pharmaceutical Company, Inc.
408 Daikanmachi, Tashiro, Tosu-shi
Saga 841

Telephone: 942-83-2101 FAX: 942-83-6119

Hoechst Japan, Ltd.
8-10-16 Akasaka, Minato-ku
Tokyo 107

Parent Company:
Hoechst AG

Telephone: 3-3479-7525
FAX: 3-3479-8552

Hokuriku Seiyaku Company, Ltd.
5-25-5 Sendagaya, Shibuya-ku
Tokyo 151

Telephone: 3-3225-1101
FAX: 3-3352-8683

ICI-Pharma Manufacturing, Ltd.
2-5-8 Imabashi, Chuo-ku
Osaka 541

Parent Company:
ICI, Ltd.

Telephone: 6-222-7000
FAX: 6-222-3655

Kaken Pharmaceutical Company, Ltd.
1 Hinode Urayasu-shi
Chiba 279

Telephone: 473-90-6111
FAX: 473-90-6160

Kayaku Company, Ltd.
2-8-16 Funado, Itabashi-ku
Tokyo 174

Telephone: 3-3558-2721
FAX: 3-3558-6458

Kissei Pharmaceutical Company, Ltd.
19-48 Yoshino
Matsumoto 399

Telephone: 263-25-9081 FAX: 263-25-7899

Kodama Ltd.
3-2 Kanda-Sakuma-cho, Chiyoda-ku
Tokyo 101

Telephone: 3-3864-8011 FAX: 3-3864-8619

Kyorin Pharmaceutical Company, Ltd.
2-5 Kanda-Surugadai, Chiyoda-ku
Tokyo 101

Telephone: 3-3293-3411 FAX: 3-3293-6588

Kyoto Pharmaceutical Industries, Ltd.
38 Nishinokyo-tsukinowa-cho, Nakagyo-ku
Kyoto 604

Telephone: 75 802 3371 FAX: 75 802 3374

Kyowa Hakko Kogyo Company, Ltd.
1-6-1 Ote-machi, Chiyoda-ku
Tokyo 100

Telephone: 3-3201-7211 FAX: 3-3284-0407

Marion Merrell Dow K.K.
2-4-1 Tsurigane-cho, Chuo-ku
Osaka 540

Parent Company:
Dow Chemical Company

Telephone: 6-941-0111 FAX: 6-946-2754

Maruho Company, Ltd.
1-6-24 Nakatsu, Kita-ku
Osaka 531

Telephone: 6-371-8876 FAX: 6-376-1724

Maruishi Pharmaceutical Company, Ltd.
2-3-5 Fushimi-machi, Chuo-ku
Osaka 541

Telephone: 6-231-6691 FAX: 6-231-6696

Maruko Seiyaku Company, Ltd
1-5-17 Kodama, Nishi-ku
Nagoya 451

Telephone: 52-524-2371 FAX: 52-531-8123

Meiji Seika Kaisha, Ltd.
2-4-16 Kyiobashi, Chuo-ku
Tokyo 104

Telephone: 3-3272-6511 FAX: 3-3271-5798

Minophagen Pharmaceutical Company
3-2-7 Yotsuya, Shinjuku-ku
Tokyo 160

Telephone: 3-3355-6563 FAX: 3-3355-6565

Mitsui Pharmaceuticals, Inc.
3-12-2 Nihonbashi, Chuo-ku
Tokyo 103

Telephone: 3-327-44711 FAX: 3-3281-4670

Mochida Pharmaceutical Company, Ltd.
1-7 Yotsuya Shinjuku-ku
Tokyo 160

Telephone: 3-3358-7211 FAX: 3-3358-7887

Nihon Schering KK
2-6-64 Nishi-miyahara, Yodogawa-ku
Osaka 532

Telephone: 6-396-2300 FAX: 6-396-6070

Nikken Chemicals Company, Ltd.
5-4-14 Tsukiji, Chuo0ku
Tokyo 104

Telephone: 3-3544-8701 FAX: 3-3545-8556

Nippon Boehringer Ingleheim Company, Ltd.
103 Aza-Takada
Kawanishi 666-01

Parent Company:
Boehringer Ingleheim AG

Telephone: 727-93-8351 FAX: 727-93-8566

Nippon Chemiphar Company, Ltd.
2-2-3 Iwamoto-cho, Chiyoda-ku
Tokyo 101

Telephone: 3-3863-1211 FAX: 3-3864-5940

Nippon Glaxo, Ltd.
3-2-9 Ginza, Chuo-Ku
Tokyo 104

Telephone: 3-3561-3181 FAX: 3-3561-1946

Nippon Kayaku Company, Ltd.
1-11-2 Fujimi, Chiyoda-ku
Tokyo 102

Telephone: 3-3237-5111 FAX: 3-3237-5091

Nippon Roche KK
3-2-3 Marunouchi, Chiyoda-ku
Tokyo 100

Telephone: 3-3214-5150 FAX: 3-3214-5388

Nippon Roussel KK
4-1-21 Nihonbashi-Muromachi, Chuo-ku
Tokyo 103

Telephone: 3-3241-7731 FAX: 3-3245-0130

Nippon Shinyaku Company, Ltd
14 Nishinosho Monguchi-cho, Kisshoin, Minami-ku
Kyoto 601

Telephone: 75-321-1111 FAX: 75-321-0678

Nippon Shoji Kaisha, Ltd.
2-2-9 Koku-machi, Chuo-ku
Osaka 540

Telephone: 6-941-0301 FAX: 6-942-6355

Nippon Zoki Pharmaceutical Company, Ltd.
2-1-2 Hirano-machi, Chuo-ku
Osaka 541

Telephone: 6-203-0441 FAX: 6-229-1127

Novo Nordisk Pharma, Ltd.
5-7 Ohdenma-cho, Nihonbashi, Chuo-ku
Tokyo 103

Telephone: 3-3249-8400 FAX: 3-3249-6228

Ono Pharmaceutical Company, Ltd.
2-1-5 Dosho-machi, Chuo-ku
Osaka 541

Telephone: 6-222-5551 FAX: 6-222-1338

Otsuka Pharmaceutical Company, Ltd
2-9 Kanda-Tsukasa-cho, Chiyoda-ku
Tokyo 101

Telephone: 3-3292-0021 FAX: 3-3295-2058

Pfizer Pharmaceuticals, Inc.
P.O. Box 226, Mitsui Bldg. Shinjuku-ku
Tokyo 163

Telephone: 3-3344-4411 FAX: 3-3348-4020

Roussel Morishita Company, Ltd.
3-3-8 Dosho-machi, Chuo-ku
Osaka 541

Telephone: 6-203-5512 FAX: 6-223-1880

Sandoz Pharmaceuticals, Ltd.
4-17-30 Nishi-Azabu, Minato-ku
Tokyo 106

Telephone: 3-3797-8000 FAX: 3-3486-6567

Sankyo Company, Ltd.
2-7-12 Ginza, Chuo-Ku
Tokyo 104

Telephone: 3-3562-0411 Fax: 3-3561-5409

Santen Pharmaceutical Company, Ltd.
3-9-19 Shimoshinjo, Higashiyodogawa-ku
Osaka 533

Telephone: 6-321-7045 FAX: 6-328-5082

Shering-Plough K.K.
1-2-6 Awaji-machi, Chuo-Ku
Osaka 541

Parent Company:
Schering-Plough, Inc.

Telephone: 6-201-1701 FAX: 6-201-1791

Searle Yakuhin K.K.
3-12-23 Kita-horie, Nishi-ku
Osaka 550

Telephone: 6-543-5511 FAX: 6-541-1362

Senju Pharmaceutical Co., Ltd.
2-5-8 Hirano-machi, Chuo-ku
Osaka 541

Telephone: 6 201 2512 FAX: 6 226 0406

Shimizu Pharmaceutical Company, Ltd.
235 Miyakami
Shimizu 424

Telephone: 543-34-1475 FAX: 543-34-2805

Shionogi & Company, Ltd.
3-1-8 Dosho-machi, Chuo-ku
Osaka 541

Telephone: 6-202-2161 FAX: 6-229-0987

Key Personnel, This Location:
Kazuo Yoshitoshi, President

SmithKline Beecham Seiyaku K.K.
6 Sanbancho, Chiyoda-ku
Tokyo 102

Parent Company:
SmithKline Beecham plc

Telephone: 3-3221-5809 FAX: 3-3264-3692

S.S. Pharmaceutical Company, Ltd.
2-12-4 Hamacho, Nihonbashi, Chuo-ku
Tokyo 103

Telephone: 3-3668-4511 FAX: 3-3667-0497

Sumitomo Pharmaceuticals Company, Ltd.
2-2-8 Dosho-machi, Chuo-ku
Osaka 541

Telephone: 6-229-8900 FAX: 6-222-2050

Taiho Pharmaceutical Company, Ltd.
1-27 Kanda-Nishiki-cho, Chiyoda-ku
Tokyo 101

Telephone: 3-3294-4527 FAX: 3-3233-4057

Taisho Pharmaceutical Company, Ltd.
3-24-1 Takada, Toshima-ku
Tokyo 171

Telephone: 3-3985-1111 FAX: 3-3985-0650

Takeda Chemical Industries, Ltd.
2-3-6 Dosho-machi, Chuo-ku
Osaka 541

Telephone: 6-204-2111 FAX: 6-204-2299

Tanabe Seiyaku Company, Ltd.
3-2-10 Dosho-mach,i Chuo-ku
Osaka 541

Telephone: 6-205-5555 FAX: 6-205-5262

Teijin, Ltd.
2-1-1 Uchisaiwai-cho, Chiyoda-ku
Tokyo 100

Telephone: 3-3506-4016 FAX: 3-3506-4498

Teikoku Hormone Manufacturing Company, Ltd.
2-5-1 Akasaka, Minato-ku
Tokyo 107

Telephone: 3-3583-8361 FAX: 3-3583-3328

Terumo Corporation
2-44-1 Hatagaya, Shibuya-ku
Tokyo 151

Telephone: 3-3374-8361 FAX: 3-3374-8399

Toa Eiyo, Ltd.
3-1-2 Kyobashi, Chuo-ku
Tokyo 104

Telephone: 3-3281-8111 FAX: 3-3272-1847

Tobishi Pharmaceuticcal Company, Ltd.
1-10-1 Yurakucho, Chiyoda-ku
Tokyo 100

Telephone: 3-3213-3771 FAX: 3-3214-4070

Tokyo Tenabe Company, Ltd.
2-2-6 Nihonbashi-Honcho, Chuo-ku
Tokyo 103

Telephone: 3-3241-5151 FAX: 3-3241-4357

Torii & Company, Ltd.
3-4-1 Nihonbashi-Honcho, Chuo-ku
Tokyo 103

Telephone: 3-3231-6811 FAX: 3-3231-6861

Toyama Chemical Company, Ltd.
3-2-5 Nishi-Shinjuku, Shinjuku-ku
Tokyo 160

Telephone: 3-3348-6611 FAX: 3-33348-6638

Tsumura & Company
12-7 Nibancho, Chiyoda-Ku
Tokyo 102

Telephone: 3-3211-0001 FAX: 3-3221-5282

Upjohn Pharmaceuticals, Ltd.
2-6-1 Nishi-Shinjuku, Shinjuku-ku
Tokyo 163

Parent Company:
The Upjohn Company

Telephone: 3-3347-8600 FAX: 3-3348-8930

Wakamoto Pharmaceutical Company, Ltd.
1-5-3 Nihonbashi-Muromachi, Chuo-ku
Tokyo 103

Telephone: 3-3279-0371 FAX: 3-3279-0393

Yamanouchi Pharmaceutical Company, Ltd,
2-3-11 Nihonbashi-Honcho, Chuo-ku
Tokyo 103

Telephone: 3-3244-3000 FAX: 3-3244-3102

Yoshitomi Pharmaceutical Industries, Ltd.
2-6-9 Hirano-machi, Chuo-ku
Osaka 541

Telephone: 6-201-1600 FAX: 6-227-5165

Zeria Pharmaceutical Company, Ltd.
10-11 Nihonbashi-Kobuna-cho, Chuo-ku
Tokyo 103

Telephone: 3-3663-2351 FAX: 3-3661-9106

---

26. KOREA

Chong Kun Dang Corp.
410 Shindorim-Dong Kuro-Gu, C.P.O. Box 3447
Seoul, 152-600

Telephone: 2 677 3841 FAX: 2 675 2570

Chunil Pharmaceutical Company, Ltd.
C.P.O. Box 6070
Seoul, 100-600

Telephone: 2 335 1001 FAX: 2 332 1001

Key Personnel, This Location:
Sang Kun Han, Chairman

Dae Woong Pharmaceutical Comapny, Ltd.
223-23 Sangdarwon-Dong
Songnam

Telephone: 342 417 7700

Dong-A Pharmaceutical Company, Ltd.
252 Yongdoo-dong Dongdasmun-ku
Seoul 130-072

Telephone: 2 920 8114 FAX: 2 924 2662

Key Personnel, This Location:
Choong-Sik Yoo, President

Dong Kook Pharmaceutical Company, Ltd.
Youngdong P.O. Box 176
Seoul 135-601

Telephone: 2 553 1161 FAX: 2 556 8542

Key Personnel, This Location:
J.H. Kang, President

Dong Wha Pharmaceutical Industrial Company, Ltd.
5 Sunwha-Dong Chung-Gu
Seoul

Telephone: 2 778 4331

Han Saem Pharma Company, Ltd.
311-1 Paekgu-R1 Paekgu-Myon
Kimje

Telephone: 65 842 2407

IL Dong Pharmaceutical Co., Ltd.
45 Shinsol-Ding Tongdaemun-Gu
Seoul

Telephone: 2 923 0081

IL Yang Pharmaceutical Industrial Co., Ltd.
24-5 Hawolgog-Dong Sung Buk-Ku
Seoul

Telephone: 2 912 5291

Jinro Pharmaceutical Company, Ltd.
231-4 Noryangjin, Dongjak-ku
Seoul 156-050

Telephone: 2 812 3101 FAX: 2 817 7083

Key Personnel, This Location:
Jae Seung Kim, President

KIC International, Ltd.
KWTC P.O. Box 171
Seoul, 136-651

Telephone: 2 554 0285 FAX: 2 556 1623

Key Personnel, This Location:
Chang Young Kim, President

Korea Green Cross Corp.
227 Kugal-Ri Kiheung-Myun
Yongin-Gun

Telephone: 3 318 2101

Korea Upjohn, Ltd.
15/F Kwanghui Bldg. 216, Chung- Gu
Seoul

Telephone: 2 275 1653

Pacific Pharmaceutical Company, Ltd.
946-16 Togok-Dong Socho-Gu
Seoul

Telephone: 2 554 4311

Pfizer Korea, Ltd.
427 Kwangjang-Dong
Seoul

Parent Company:
Pfizer Corp.

Key Personnel, This Location:
Kim Joong Bae, President

Samik Pharmaceutical Company, Ltd.
Dongdaemun P.O. Box 5
Seoul 130600

Telephone: 2 928 0661 FAX: 2 924 3749

Key Personnel, This Location:
Sae Young Lee, President

Sam Sung Pharmaceutical Industrial Company, Ltd.
199-6 Mojin-Dong Sungdong-Gu
Seoul

Telephone: 2 457 1241

Sang-A Pharmaceutical Company, Ltd.
640-9 Tungchon-Dong
Seoul

Telephone: 2 694 6511

Su Heung Capsule Company, Ltd.
317 Songnae-Dong
Puchon

Telephone: 32 628 841

Taejoon Pharmaceutical Company, Ltd.
657-87 Hannam 1-Dong
Seoul

Telephone: 2 798 6601

Taikwang Sanofi Pharmaceutical Company, Ltd.
778-3 Wonshi-Dong Ansan-Shi
Kyonggi-Do, Kyo

Telephone: 839 732

Yeongdong Pharmaceutical Corp.
288-14 Singdang-Dong, Jung-Gu
Seoul 100456

Telephone: 2 7233 3531 FAX: 2 7233 5687

---

27. MALAYSIA

Behn Meyer & Company (Malaysia) Sdn. Bhd.
P.O. Box 102, Jalan Sultan
16750 Petaling Jaya, Selengor

Telephone: 3 756 1233 FAX: 3 756 2118

Camden Industries (Malaysia), Sdn. Bhd.
998 Solok Perusahaan Tiga
Kawasan MIEL, Prai Industrial Estate,
13600 Prai, Penang

Telephone: 4 307 319 FAX: 4 397 234

Key Personnel, This Location:
Heah Chew Teng, Managing Director

Duopharma (Malaysia), Sdn. Bhd.
27 & 29 Jalan 16 Kawasan 3
Taman Klang Jaya, 41200 Klang

Telephone: 3 323 2759 FAX: 3 323 3923

Key Personnel, This Location:
Cheah Ting Poh, Managing Director

Eagle & Pagoda Brand Teck Aun Medical Factory, Sdn. Bhd.
486-489 Jalan Sungai Rambai
14000 Bukit Mertajam, Province Wellesley
Penang

Telephone: 4 511 533 FAX: 4 593 675

Key Personnel, This Location:
Oon Prek Seng, Managing Director

The East Asiatic Company (Malaysia), Sdn. Bhd.
Berhad-Dumex Division
1 Jalan 205
46050 Petaling Jaya, Selangor

Telephone: 3 791 3322 FAX: 3 791 3561

Key Personnel, This Location:
Ken Ong, Executive Director

Glaxo Malaysia, Sdn. Bhd.
Lot 3, Jalan 51A/221
46700 Petaling Jaya,Selangor

Parent Company:
Glaxo Group, plc

Telephone: 3 755 2633 FAX: 3 755 6830

Key Personnel, This Location:
Zainal Abidin Majid, Managing Director

Malayan Pharmaceutical Factory, Sdn. Bhd.
3 Jalan Semangat
46200 Petaling Jaya, Selangor

Telephone: 3 756 6354 FAX: 3 756 0066

Key Personnel, This Location:
Mah Yeat Chin, Sales Manager

PharmaMalaysia, Sdn. Bhd.
69 & 71 Medan Setia 1
Bukit Damansara
50490 Kuala Lumpur

Telephone: 3 255 3166 FAX: 3 255 4971

Key Personnel, This Location:
C.B. Lee, Manager

Procter & Gamble Malaysia, Sdn. Bhd.
Lot 16B Jalan Section 51a
Petaling Jaya, Selangor

Parent Company:
Procter & Gamble Company

Telephone: 3 774 6933

Sandoz Products (Malaysia) Sdn. Bhd.
No. 35 1st Floor, The Highway Centre
Jalan 51/205
46050 Petaling Jaya

Telephone: 3 791 6784 FAX: 3 791 6819

Key Personnel, This Location:
Lim Sow Mun, Deputy Managing Director

Upha Corp. (Malaysia) Sdn. Bhd.
Lot 2 & 4, Jalan P/7 Section 13
Bangi Industrial Estate, Bandar Baru Bangi
43000 Kajang, Selangor

Telephone: 3 825 9922 FAX: 3 825 7930

Key Personnel, This Location:
Chong Fong Choo, Sales Manager

---

28. MEXICO

Abbott Laroratories
De Mexico S.A. de C.V.
Av Coyoacan 1622
México City, D.F. 03100

Telephone: 5 534 3680

Parent Company:
Abbott Laboratories

Andre Bigaux S.A. de C.V.
Ave. Vincente Garcia Torres 235
México City, D.F. 33000

Arancia S.A. de C.V.
Lopez Cotilla 2030
Guadalajara, Jalisco 10000

Telephone: 3 651 1212

Armstrong Laboratories De Mexico S.A. de C.V.
Division Del Norte No. 3311
México City, D.F. 38000

Telephone: 5 618 4875

Baxter S.A. de C.V.
Oklahoma No. 14-3er Piso
México City, D.F. 03810

Telephone: 5 669 1122

Bayer De Mexico S.A. de C.V.
Bloulevard Miguel De
Cervantes Saavedra 259
México City, D.F. 11520

Telephone: 5 250 2055

Beecham Farmaceutica S.A. de C.V.
Miguel Angel De Quevedo 307
México City, D.F. 04360

Telephone: 5 554 0911

BYK Gulden S.A. de C.V.
10 De Mayo No. 130
Naucalpan De Juarez,
México City, D.F. 53500

Telephone: 5 358 4347

Ciba Geigy Mexicana S.A. de C.V.
Calz De Tlalpan 3058
México City, D.F. 04850

Telephone: 5 677 1088

Farmitalia Carlo Erba S.A. de C.V.
Miguel A. de Quevedo 555
Mexico City, D.F. 04310

Telephone: 5 659 8246 FAX: 5 554 8828

Key Personnel, This Location:
Enrico Soave Bianchini, Director General

Fermic S.A. de C.V.
Reforma No. 873
Mexico City, D.F. 09850

Telephone: 5 656 6498 FAX: 5 656 1542

Fisons de Mexico S.A. de C.V.
Calz De Las Armas 110
Tlalnepantla, México City, D.F. 04090

Parent Company:
Fisons plc

Telephone: 5 394 7199

Grupo Roussel S.A. de C.V.
Av Universidad 1738
México City, D.F. 04000

Telephone: 5 658 6377

Janssen Farmaceutica S.A. de C.V.
Canoa 79
México City, D.F. 01090

Telephone: 5 683 1177

Laboratorios Bristol De Mexico S.A. de C.V.
1o de Mayo 226 Antes Av. Walte
Naucalpan De Juarez, 05000

Telephone: 5 358 5344

Laboratorios Colombia S.A. de C.V.
Calz Del Hueso 670
México City, D.F. 04000

Telephone: 5 677 7722

Laboratorios Dermatologicos
Barranca del Muerto 182
México City, D.F. 01600

Telephone: 5 660 7294 FAX: 5 660 6900

Key Personnel, This Location:
Fernando Ahumada Ayala, Director General

Laboratorios Pisa S.A. de C.V.
Av. Espana 1840
Guadalajara, Jalisco 44190

Telephone: 3 610 1935 FAX: 3 612 2989

Key Personnel, This Location:
Jose Ignacio Rodriguez Ramirez, Director

Laboratorios Promeco de Mexico
Insurgentes Sur 1457, Piso 8
México City, D.F. 03900

Telephone: 5 598 3866 FAX: 5 598 3867

Key Personnel, This Location:
Francisco Perez Atamoros, Director

Laboratorios Sensorian S.A. de C.V.
Andres bello 45-20o Y 21 Pisos
México City, D.F. 05600

Telephone: 5 280 4999

Laboratorios Silanes S.A. de C.V.
Amores 1304
México City, D.F. 01000

Telephone: 5 604 1994

Laboratorios Sofia S.A. de C.V.
Av. Hidalgo 737
Guadalajara, Jalisco 10000

Telephone: 3 613 5500

Laboratorios Terrier S.A. de C.V.
Lago Alberto 369
México City, D.F. 03200

Telephone: 5 250 3411

Eli Lilly y Cia. de Mexico S.A. de C.V.
Calz De Tlalpan 2024
México City, D.F. 04220

Telephone: 5 549 3140

Pfizer S.A. de C.V.
Damas 120
México City, D.F. 03900

Parent Company:
Pfizer Corp.

Telephone: 5 660 3420

Quimica Hoechst de Mexico S.A. de C.V.
Tecoyotitla 412
México City, D.F. 01050

Telephone: 5 228 9000 FAX: 5 661 3792

Key Personnel, This Location:
Guenter Martin, President

Quimica Knoll de Mexico S.A. de C.V.
Candeleria No. 186
México City, D.F. 04330

Telephone: 5 549 3740

Rhône Poulenc Pharma De Mexico S.A. de C.V.
Division Del Norte 3315
México City, D.F. 04380

Rhône-Poulenc Rorer S.A. de C.V.
Matias Romero No. 216
México City, D.F. 03100

Richardson Vicks S.A. de C.V.
Loma Florida No. 32
México City, D.F. 05100

Key Personnel, This Location:
Luis F. Velez, President

A. H. Robins de Mexico S.A. de C.V.
Autopista Mexico Queretaro, Km 375
Cuautitlán, Mexico 54730

Sanofi de Mexico S.A. de C.V.
Calz Mexico-Xochimilco 4832
México City, D.F. 14370

Searl de Mexico S.A. de C.V.
Calz Del Hueso 859
México City, D.F. 14300

Schering-Plough S.A. de C.V.
Av. 16 De Septembre No. 301
México City, D.F. 16090

---

29. MONACO

Laboratories Theramex
4/6 Av Prince Hereditaire Albert
Monaco 98007

Telephone: 93 050 808

---

30. NETHERLANDS

Affymax N.V.
Drentestraat 20
Amsterdam, 1083 HK

Telephone: 20 644 2590

Key Personnel, This Location:
J.D. Diekman, President
K.J. Nussbacker, Exec. Vice President

Akzo N.V.
Velperweg 76, Box 9300
Arnham, 6800 SB

Telephone: 85 664 433

Key Personnel, This Location:
A.A. Loudon, Chairman

Bufa B.V.
Molenwerf 13
Uitgeest 1911 DB

Telephone: 2513 191 00 FAX: 2513 159 60

Emonta B.V.
P.O. Box 197
Drachten, 9200 AD

Telephone: 51 201 8985 FAX: 51 201 3982

Key Personnel, This Location:
John Houthoff, Manager

Gist-Brocades B.V.
Wateringsweg 1, Delft
Zuid-Holla, 2600 MA

Telephone: 15 799 111

ICN Pharmaceuticals Holland B.V.
P.O. Box 328
Zoetermeer, 2723 RM

Telephone: 79 412 121 FAX: 79 421 208

Key Personnel, This Location:
F.A. Hijmans, Manager

Solvay Duphar B.V.
C. J. Van Houtenlaan 36, Weesp
Utrecht, 1380 AA

Telephone: 29 407 7000

---

31. NIGERIA

Wellcome Nigeria, Ltd.
Oba Akran Ave.
Ikeja, Lagos

Parent Company:
Wellcome Foundation, Ltd.

32. NORWAY

Apothekernes Laboratorium AS
P.O. Box 158, Skoyen
Oslo N-0212

Telephone: 22 52 90 00 FAX: 22 50 58 60

Collett Marwell Hauge AS
Drammensvelen 852
Asker, Akershus N-1371

Nygaard & Co. AS Nyco
Nycoveien 2
Oslo, Akershus

Telephone: 22 17 14 90

---

33. PAKISTAN

Abbott Laboratories Pakistan, Ltd.
Opp Radio Pakistan
Transmission Centre Landhi
Karachi, Sind 22

Telephone: 21 330 045

Parent Company:
Abbott Laboratories

Fazal Din & Sons (Pvt.), Ltd.
53 Shahrah-e-Quaid-e-Azam
Lahore

Telephone: 42 723 1762 FAX: 42 723 1770

Key Personnel, This Location:
Mumtaz Sheikh Sheikh, Chairman

Glaxo Laboratories (Pakistan), Ltd.
3f Press Trust House 1, 1 Chundrigar Rd.
Karachi, Sind 1

Telephone: 21 210 561

Lahore Chemical & Pharmaceutical
137 Ferozpur Road
Lahore 54600

Telephone: 42 758 8628 FAX: 42 723 1770

Key Personnel, This Location:
Ijaz A Mumtaz, Managing Director

Reko Pharmaceuticals (Pvt.), Ltd.
P.O. Box 1129
Lahore

Telephone: 42 522 1201 FAX: 42 666 1981

Key Personnel, This Location:
Mian Khalid Mansur, Managing Director

Sandoz Pakistan, Ltd.
6 th Floor, Mahboob Chambers
Abdullah Haroon Rd.
Karachi, Sind

Telephone: 21 292 033

Wellcome Pakistan, Ltd.
F/26B Site
Karachi, Sind

Telephone: 21 290 515

---

35. PARAGUAY

Scavone Y Cia Vicente SACI
Pte. Franco 570
Asuncion

---

35. PERU

Abbott Laboratorios S.A.
Av. Brasil 2370
Pueblo Libre, Lima 21

Telephone: 14 614 791

Key Personnel, This Location:
Nelson Pizarro

Parent Company:
Abbott Laboratories

American Farmaceutica del Peru S.A.
Vargas Machuca 481
Miraflores, Lima 18

Telephone: 14 456 217

Key Personnel, This Location:
Jose Diez Canseco

Bristol-Myers Peruana S.A.
Av. Republica de Panama 4575
Surquillo, Lima 34

Telephone: 14 454 040

Key Personnel, This Location:
Jose Freyre

CODIEX S.A.
Carretera Central Km. 4.700
Ate, Lima 03

Telephone: 14 359 141

Key Personnel, This Location:
Jose Gordillo

Corporacion Farmaceutica S.A.
Avenue Paseo de la Republica 3074, Oficina 802
San Isidro, Lima 27

Telephone: 14 4410 299

Key Personnel, This Location:
Manuel Podesta

Farmaceutica del Pacifico S.A.
Republica de Panama 4825
Surquillo, Lima 34

Telephone: 14 451 540

Key Personnel, This Location:
Arturo Mongurt

Hoechst Peruana S.A.
Carratera Centra, Km 3.700
Ate, Lima 03

Telephone: 14 310 505

Key Personnel, This Location:
Wolf Derfler

Laboratorio Farmaceutico Peruano S.A.
Av. Republic de Panama 1773
La Victoria, Lima 13

Telephone: 14 720 497

Key Personnel, This Location:
Fernando Stahl

Laboratorios Alfa S.A.
Av. Colonial 1560
Lima, Lima 01

Telephone: 14 375 813

Key Personnel, This Location:
Charles Wetzel

Laboratorios Farmaceuticos Markos
Av. Separadora Industrial, Mz-I, Lote 13
Ate, Lima 03

Telephone: 14 360 307

Key Personnel, This Location:
Marcelino Orihuel

Laboratorios Farmindustria S.A.
Mariscal Miller 2151
Lince, Lima 14

Telephone: 14 701 111

Key Personnel, This Location:
Vittorio Azzariti De Mentri, General Manager

Laboratorios Trifarma S.A.
Santa Rosa y Santa Maria S/n
Ate-Vitarte, Lima 03

Telephone: 14 740 696

Key Personnel, This Location:
Luis Ormeno

Eli Lilly Interamerica, Inc.
Las Begonias 451
San Isidro, Lima 27

Parent Company:
Eli Lilly & Company

Telephone: 14 422 626

Key Personnel, This Location:
Jorge Milberg

Madfarma S.A.
Av. Francisco Pizarro 345-2do. Piso
Rimac, Lima 25

Telephone: 14 814 189

Key Personnel, This Location:
Alfonso Huamani

Upjohn Inter-America Corp.- Peru
Calle 32 No. 432 Urb. Corpac.
San Isidro, Lima 27

Parent Company:
The Upjohn Company

Telephone: 14 750 641

Key Personnel, This Location:
Luciano Ampuero

---

36. PHILIPPINES

Nicholas Kiwi Philippines, Inc.
329 Makati Metro Manila
Makati, M.M. 1200

Telephone: 2 818 0616 FAX: 2 815 0513

E.R. Squibb & Sons
Philippines Corp.
Pasong Tamo Cornermalugay Sts.
Makati, Metro Manilla

Telephone: 2 880 6666

Warner -Lambert Philippines, Inc.
National Life Insurance Bldg.
San Juan, Metro Manilla

Parent Company:
Warner-Lambert Company

Telephone: 2 721 5251

Key Personnel, This Location:
Honorato Timbol, Sales Manager

---

37. POLAND

Grodziskie Zaklady Farmaceutyczne Polfa
ul. Poniatowskiego 5
05-700 Grodzisk Maz.

Telephone: 22 552 041

Jeleniogorskie Zaklady Farmaceutyczne Polfa
ul. Wincentego Pola 13/18
58-500 Jelenia Gora

Telephone: 75 228 21 FAX: 75 244 45

Krakowskie Zaklady Farmaceutyczne Polfa
ul. Mogilska 80
31-546 Krakow

Telephone: 12 117 744 FAX: 12 111 047

Key Personnel, This Location:
Edward Sorowka, Managing Director

Kutnowskie Zaklady Farmaceutyczne Polfa
ul. Sienkiewicza 25
99-300 Kutno

Telephone: 218 331 25 FAX: 218 423 74

Lodzkie Zaklady Farmaceutyczne Polfa
ul. Drewnowska 43/47
91-002 LOdz

Telephone: 42 579 000

Lowickie Zaklady Farmaceutyczne Polfa
ul. Wolnosci 2
99-420 Lyszkowice k. Lowicza

Telephone: 201 61 75

Lubelskie Zaklady Farmaceutyczne Polfa
ul. Boczna Lubomelskiej 13
20-070 Lublin

Telephone: 81 272 22

Poznanskie Zaklady Farmaceutyczne Polfa
ul. Grunwaldzka 189
60-322 Poznan

Telephone: 61 672 750

Starogardzkie Zaklady Farmaceutyczne Polfa
ul. Pelpinska 19
83-200 Starogard Gdanski

Telephone: 69 234 01 FAX: 69 223 53

Warszawskie Zaklady Farmaceutyczne Polfa
ul. Karolkowa 22
01-207 Warsaw

Telephone: 22 322 326

---

38. PORTUGAL

Laboratorios Astral S.A.
Avenida Gomes Pereira 74-76
Lisbon 1500

Ortomaia de Jose de A. Ribeiro
Apart 31, Aeroporto do Porto
Porto, 4470

Telephone: 2 498 1532 FAX: 2 941 1470

---

39. RUSSIA

Pharmimex
Pushkinskya Str., 7
101668, Moscow

Telephone: (095) 292 4848 FAX: (095) 292 9889

Key Personnel, This Location:
Alexander A. Apazov, General Director

Rospharmatsia
1st Kolobovsky Per., 19
103516, Moscow

Telephone: (095) 299 4225 FAX: (095) 299 8034

Key Personnel, This Location:
Inna N. Laskina, General Director

---

40. SAUDI ARABIA

Bristol-Myers Squibb
P.O. Box 16760
Jeddah, 21474

Parent Company:
Bristol-Myers Squibb Company

Telephone: 2 643 6470

FAX: 2 643 8428

Saudi Pharmaceutical Industries & Medical Appliances Corp.
7th Floor, Akariya Bldg. Sitte
Riyadh 11455

Telephone: 1 477 4481

FAX: 1 477 3961

Key Personnel, This Location:
Abdullah Abdulkadar, Managing Director

---

41. SINGAPORE

Beacons Chemicals (Pte.), Ltd
53 Quality Road, Jurong Town
Singapore 2261

Telephone: 265 2336

Key Personnel, This Location:
David A.H. Lam, Managing Director

Beecham Pharmaceuticals (Pte.), Ltd.
Quality Road, Jurong Town
Singapore 2261

Telephone: 265 3711

Key Personnel, This Location:
J.C. Sanderson, Managing Director

Drug Houses Of Austrailia (Asia) Pte., Ltd.
2 Chia Ping Road, # 09-01
Haw Par, Tiger Balm Building
Singapore 2261

Telephone: 265 2777

Key Personnel, This Location:
Thia Siow Khee, General Manager

Glaxochem (Pte.), Ltd.
1 Pioneer Sector 1, Jurong Town
Singapore 2262

Telephone: 861 2111 FAX: 861 1806

Key Personnel, This Location:
R. Costello, Managing Director

Innovation Pharmaceutical Industries Pte., Ltd.
204 Pandan Loop, Jurong Town
Singapore 0512

Telephone: 778 5415

Key Personnel, This Location:
Lee Kam Chi, Managing Director

Little Pharmaceuticals Suppliers Pte., Ltd.
998 #005 Toa Paych, North Flatted Factory
Singapore 1231

Telephone: 252 1331

Key Personnel, This Location:
Lau Guan Kim, Managing Director

Stiefel Laboratories Pte., Ltd.
103 Gul Circle, Jurong Town
Singapore 2262

Telephone: 862 1333

Key Personnel, This Location:
Robert Caiola, Operations Manager

---

42. SPAIN

Allergan S.A.E
Avda. de la Industria, 24 Poligono
Madrid, ES 18037

Telephone: 1 807 6310 FAX: 1 803 6200

Antibioticos Farma S.A.
Antonio Lopez 109-111
Madrid, Madrid 28026

ASAC Pharmaceutical International
Sagitario 12
Alicante, ES 03006

Telephone: 6 528 6700

Key Personnel, This Location:
Eliseo Q. Almagro, Director General

Boehringer Ingleheim Espana S.A.
Pablo Alcover 31-33
Barcelona, Barcelona 08017

Ciba Geigy S.A.
Balmes 117
Barcelona, Barcelona 08008

Divasa-Farmavic S.A.
Ctra. Sant Hispolit, Km 71
Grup, ES 08518

Telephone: 3 866 0100 FAX: 3 889 0131

Key Personnel, This Location:
Josep Jutglar, Manager

Igoda S.A.
Ctra. Nacional 152, Km 19.5
Mollet Del Valles, Barcelona 08100

Telephone: 3 570 5750

Laboratorios Bama-Geve S.A.
Av. Diagonal 456, 4a Planta
Barcelona, ES 08006

Telephone: 3 415 4822 FAX: 3 415 3792

Key Personnel, This Location:
Germano Natali, Manager

Laboratorios Cinfa S.A.
Olaz-Chipi 10
Huarte-Pamplona, ES 31620

Telephone: 48 330 763 FAX:48 330 367

Key Personnel, This Location:
Javier Ibilcieta, Managing Director

Laboratorios Grifols S.A.
Jesus Y Maria 6
Barcelona, Barcelona 08022

Telephone: 3 571 0100

Laboratorios Menarini S.A.
Alfonso XII 587
Barcelona, Barcelona 08912

Telephone: 3 387 5600

Quimica Farmaceutica Bayer S.A.
Calabria 268 Esquina a Paris
Barcelona, Barcelona 08029

Telephone: 3 430 9600

Sandoz Pharma S.A.E.
Gran Via De Les Corts, Catalanes 764
Barcelona 08013

Telephone: 3 245 1700

Searle Iberica S.A.
C. de la Granja 23
Alcobendas, ES 28033

Telephone: 1 637 7100

Key Personnel, This Location:
Jose Noguera, Director

---

43. SWEDEN

Allergon AB
Välinge 2090
262 92 Ängelholm

Telephone: 42 692 40 FAX: 42 691 17

Apoteksbolaget AB
Humleårdsg 20
105 14 Stockholm

Telephone: 8 666 7000 FAX: 8 666 7515

Armofin AB
Verkstadsv 2, Box 223
241 23 Eslöv

Telephone: 413 170 15 FAX: 413 609 75

Astra AB
Kvambergagatan 16, Sodertalje
151 85 Stockholm

Telephone: 8 553 2980

Biocomplex AB
Malmv 46, Box 623
191 26 Sollentuna

Telephone: 8 35 83 37 FAX: 8 96 97 71

Biogram AB
Borrg 31 A, Box 50577
201 15 Malmö

Telephone: 40 28 70 90 FAX: 40 29 26 83

Bröste AB
Aminog 34
431 53 Mölndal

Telephone: 31 776 95 95 FAX: 31 776 96 76

DCG Farmaceutiska AB
Fältspatsg 3, Box 2123
421 02 Västra Frölunda

Telephone: 31 47 30 11 FAX: 31 47 88 31

Farmitalia Carlo Erba AB
Alv 5 Box 3511
183 03 Täby

Telephone: 8 756 7085 FAX: 8 756 8247

Fisons Sweden AB
Oxholmsgränd 3, Box 170
127 23 Skärholmen

Telephone: 8 710 00 20 FAX: 8 740 39 96

Kabi Pharmacia Sverige AB
Rapsg 7
751 82 Uppsala

Telephone: 18 16 3000 FAX: 18 12 6077

Nobelpharma Produktion AB
Dimbov 2
691 51 Karlskoga

Telephone: 586 818 50 FAX: 586 365 60

Perstorp Pharma
Scheelev 17
223 70 Lund

Telephone: 46 18 55 10 FAX: 46 13 76 70

Pharmacia AB
Rapsgatan 7
751 82 Uppsala

Telephone: 18 16 3000 FAX: 18 12 6077

Pfizer AB
Nytorpsv 36, Box 501
183 03 Täby

Telephone: 8 758 01 30 FAX: 8 768 37 37

Rhône-Poulenc Sevrige AB
Folkungag 122, Box 4189
102 64 Stockholm

Telephone: 8 23 7480 FAX: 8 641 8863

Simplus AB
Tallv 5
792 52 Mora

Telephone: 250 162 30 FAX: 250 162 20

Tricum AB
Sporthallsv 6, Box 14
263 21 Höganäs

Telephone: 42 33 1115 FAX: 42 33 2209

---

44. SWITZERLAND

Ciba-Geigy AG
Klybeckstrasse 141
Basel, Basel Stad 4057

Telephone: 61 697 1111

Galencia Vertretungen AG
Untermattweg 8
Bern, Bern 3027

Telephone: 31 552 222

Hoffman LaRoche AG / Roche Holdings AG
Grenzacherstrasse 124
Basel 4002

Telephone: 61 668 11 11

Key Personnel, This Location:
Fritz Gerber, President
Jurgen Drews, R&D Director

Psori–Med AG
Dufourstrasse 101
Zurich, ZH 8034

Telephone: 1 383 3438 FAX: 1 383 0360

Key Personnel, This Location:
Hans–Joerg Lutz, Chairman

Sandoz Pharma AG
Lichstrasse 35
Basel, Basel Stad 4002

Telephone: 61 344 1111

Key Personnel, This Location:
Marc Moret, Chairman
Hans Letsch, Vice Chairman

Sygena, Ltd.
Eichenweg 1
Liestal 4410

Telephone: 61 901 5959 FAX: 61 901 6152

Zellweger Uster, Ltd.
Wilstrasse 11
Uster, ZH 8610

Telephone: 1 943 2211 FAX: 1 940 7079

Key Personnel, This Location:
M. Meili, Director

---

45. SYRIA

The Arabian Medical Company
Eastern Ghouta
Damascus

Telephone: 431 073

---

46. TURKEY

Abbott Laboratories
Dr. F. Kerim Gokay Caddesi 31/2
81190 Altunizada, Istanbul

Parent Company:
Abbott Laboratories

Telephone: 216 325 5310 FAX: 216 325 4770

Key Personnel, This Location:
Roberto Reyes, Managing Director

Johnson & Johnson
Sihhi Malzeme, Ebulula Caddesi 23
80620 1. Levent, Istanbul

Telephone: 212 280 6903 FAX: 212 280 6955

Key Personnel, This Location:
John Ross-Smith, General Manager

Pfizer Ilaclari A.S.
Muallim Naci Caddesi-Bostan Cikmazi 3
80840 Ortakoy, Istanbul

Telephone: 212 260 2210 FAX: 212 258 0624

Key Personnel, This Location:
Alev Sonat, Manager

Wyeth Laboratuarlari A.S.
Maya Akar Center, Buyukdere Caddesi 110/102 - Kat 9
80280 Esentepe, Istanbul

Telephone: 212 211 7612 FAX: 212 211 7616

---

47. UKRAINE

Combio, Ltd.
9 Leontovicha St.
Kiev 252035

Telephone: (044) 516 2420 FAX: (044) 516 2422

Key Personnel, This Location:
Georgiy Leonidovick Volkov, Director

Darnitsa
13 Borispilska St.
Kiev 253093

Telephone: (044) 556 0072 FAX: (044) 556 7646

Key Personnel, This Location:
Vladimir Antonovich Zagoriy, General Director

Dnipropetrovsk Chemical And Pharmaceutical Plant
12 Prospect K. Marska
Dnipropetrovsk 320095

Telephone: (0562) 46 0415 FAX: (0562) 46 5066

Galichpharm
6/8 Opryshkovska St.
Lviv 290601

Telephone: (0322) 52 1929 FAX: (0322) 52 8498

Kievmedpreparat
139 Saksahanskoho St.
Kiev 252032

Telephone: (044) 221 1903 FAX: (044) 216 9645

Key Personnel, This Location:
Vladimir Vasilyevich Vinohradov, Director

Lekhim
18a Ostrovskoho St.
Kiev-35 252035

Telephone: (044) 244 0246 FAX: (044) 276 9126

Key Personnel, This Location:
Valeriy Kostiantynovich Pechaev, Director

Luhansk Chemical And Pharmaceutical Plant
17 Kirov St.
Luhansk 348023

Telephone: (0642) 52 5531

Pharmac
74 Vul. Frunze
Kiev 254072

Telephone: (044) 417 1055 FAX: (044) 417 6049

Key Personnel, This Location:
Ivan Ivanovich Prybudchenko, General Director

State Center of Medicaments
33 Vul. Anatomicheskaya
Kharkiv 310085

Telephone: (0572) 44 1033 FAX: (0572) 44 1118

Key Personnel, This Location:
Victor Petrovich Georgievsky, Director

Zdorovye Pharmaceutical Company
22 Shevchenka St.
Kharkiv 310013

Telephone: (0572) 47 1653

---

48. UNITED KINGDOM

Abbott Laboratories, Ltd.
Abbott House, Norden Road
Maidenhead, Berks SL6 4XE

Parent Company:
Abbott Laboratories

Telephone: 628 773 355 FAX: 628 644 305

Allen & Hanburys, Ltd.
Stockley Park West
Uxbridge, Middlesex UB11 1BT

Telephone: 81 990 9888 FAX: 81 990 4375

Allergan, Ltd.
Coronation Road
High Wycombe, Bucks HP12 3SH

Telephone: 494 444 722 FAX: 494 473 593

Amgen, Ltd.
240 Cambridge Science Park
Milton Road, Cambridge CB4 4WD

Telephone: 223 420 305 FAX: 223 423 049

Armour Pharmaceutical Company, Ltd.
RPR House, 52 St. Leonards Road
Eastbourne, East Sussex BN21 3YG

Telephone: 323 410 200 FAX: 323 410 306

Astra Pharmaceuticals, Ltd.
Home Park
Kings Langley, Herts WD4 8DH

Telephone: 923 266 191 FAX: 923 260 431

Bayer, plc, Pharmaceutical Division
Bayer House, Strawberry Hill
Newbury, Berkshire RG13 1JA

Telephone: 635 563 371 FAX: 635 563 357

Beecham Group plc
4 New Horizons Court, Harlequin Avenue
Brentford, Middlesex TW8 9EP

Telephone: 81 975 2000

Berk Pharmaceuticals
Brampton Road, Hampden Park
Eastbourne, East Sussex BN22 9AG

Telephone: 323 501 111 FAX: 323 502 275

Bristol-Myers Squibb Pharmaceuticals, Ltd.
141/149 Staines Road
Hounslow, Middlesex TW3 3JA

Parent Company:
Bristol-Myers Squibb Co.

Telephone: 81 572 7422 FAX: 81 577 1756

Ciba Pharmaceuticals
Wimblehurst Road
Horsham, West Sussex RH12 4AB

Telephone: 403 272 827 FAX: 403 323 053

Cilag, Ltd.
P.O. Box 79, Saunderton
High Wycombe, Bucks HP14 4HJ

Telephone: 494 563 541 FAX: 494 565 372

CP Pharmaceuticals, Ltd.
Ash Road North
Wrexham, Clwyd LL13 9UF

Telephone: 978 661 261 FAX: 978 660 130

Dista Products, Ltd.
Dextra Court, Chapel Hill
Basingstoke, Hampshire RG21 2SY

Telephone: 256 473 241 FAX: 256 485 858

DuPont Pharmaceuticals, Ltd.
Avenue One
Letchworth Garden City, Herts SG6 2HU

Telephone: 462 482 648 FAX: 642 488 206

Fisons plc - Pharmaceutical Division
12 Derby Road
Loughborough, Leichs LE11 0BB

Telephone: 509 611 001 FAX: 509 611 116

Genzyme Pharmaceuticals and Fine Chemicals
Hollands Road
Haverhill, Suffolk CB9 8PU

Parent Company:
Genzyme Corp.

Telephone: 440 703 522 FAX: 440 707 783

Glaxo Wellcome Pharmaceuticals UK, Ltd.
Stockley Park West
Uxbridge, Middlesex UB11 1BT

Parent Company:
Glaxo Group plc

Telephone: 81 990 9000 FAX: 81 990 4321

Key Personnel, This Location
Sean P. Lance, Director

Immuno, Ltd.
Arctic House, Rye Lane, Dunton Green
Nr. Sevenoaks, Kent TN14 5HB

Telephone: 732 458 101 FAX: 732 743 477

Hoechst Roussel, Ltd.
Broadwater Park, Denham
Uxbridge, Middlesex UB9 5HP

Telephone: 895 834 343 FAX: 895 834 479

Janssen Pharmaceutical, Ltd.
Grove
Wantage, Oxon OX12 0DQ

Telephone: 235 777 333 FAX: 235 777 444

Knoll, Ltd.
Fleming House, 71 King Street
Maidenhead, Berks SL6 1DU

Parent Company:
BASF AG

Telephone: 628 776 360 FAX: 628 776 579

Lilly Industries, Ltd.
Dextra Court, Chapel Hill
Basingstoke, Hampshire RG21 2SY

Parent Company:
Eli Lilly Group, Ltd.

Telephone: 256 473 241 FAX: 256 485 858

3M Health Care, Ltd.
3M House, Morley Street
Loughborough,
Leicestershire, LE11 1EP

Parent Company:
3M Company

Telephone: 509 611 611 FAX: 509 237 288

Marion Merrell Dow, Ltd.
Lakeside House, Stockley Park
Uxbridge, Middlesex UB11 1BE

Telephone: 81 848 3456 FAX: 81 848 5404

May & Baker
Rainham Road South
Dagenham, Essex RM10 7XS

Telephone: 81 919 3060 FAX: 81 919 2140

Medeva, plc
10 Saint James's Street
London SW1A 1EF

Telephone: 71 839 3888 FAX: 71 930 1514

Key Personnel, This Location:
B. D. Taylor, Chairman
William W. Gerard, Vice Chairman

Merck Sharp & Dohme, Ltd.
Hertford Rd.
Hoddesdon, Herts EN11 9BU

Telephone: 922 467 272 FAX: 922 468 175

Merieux UK, Ltd.
Clivemont House, Clivemont Road
Maidenhead, Berks SL6 7BU

Telephone: 628 785 291 FAX: 628 717 22

Norton Healthcare, Ltd.
Gemini House, Flex Meadow
Harlow, Essex CM19 5TJ

Telephone: 279 426 666 FAX: 279 432 110

Organon Laboratories, Ltd.
Cambridge Science Park, Milton Road
Cambridge CB4 4FL

Telephone: 223 423 445 FAX: 223 424 368

Parke Davis & Co., Ltd.
Lambert Court, Chestnut Avenue
Eastleigh, Hampshire SO53 3ZQ

Parent Company:
Warner-Lambert Co., Ltd.

Telephone: 703 620 500 FAX: 703 614 604

Pfizer, Ltd.
Ramsgate Road
Sandwich, Kent CT13 9NJ

Parent Company:
Pfizer Group, Ltd.

Telephone: 304 616 161 FAX: 304 616 221

Procter & Gamble Pharmaceuticals UK, Ltd
Lovett House, Lovett Road
Staines, Middlesex TW18 3AZ

Parent Company:
Procter & Gamble Company

Telephone: 784 495 000 FAX: 784 495 297

Reckitt & Colman, plc.
One Burlington Lane
Chiswick, London W4 2RW

Telephone: 81 994 6464

Key Personnel, This Location:
M.J. Colman, Chairman
V.L. Sankey, CEO

Rhône-Poulenc Rorer, Ltd.
RPR House, 52 St. Leonards Rd.
Eastbourne, East Sussex BN21 3GY

Parent Company:
Rhône-Poulenc Rorer, Inc.

Telephone: 323 534 000 FAX: 323 534 080

A.H. Robins Company, Ltd.
Huntercombe Lane South, Taplow
Maidenhead, Berks SL6 0PH

Telephone: 628 604 377 FAX: 628 666 368

Roche Products, Ltd.
40 Broadwater Road
Welwyn Garden City, Herts AL7 3AY

Telephone: 707 366 000 FAX: 707 338 297

Sandoz Pharmaceuticals (UK), Ltd.
Frimley Business Park, Frimley
Camberly, Surry GU16 5SG

Telephone: 276 692 255 FAX: 276 692 508

Sanofi Winthrop, Ltd.
One Onslow Street
Guildford, Surrey GU16 4YS

Telephone: 483 505 515 FAX: 483 354 320

Schering-Plough, Ltd.
Schering-Plough House, Shire Park
Welwyn Garden City, Herts AL7 1TW

Telephone: 707 363 636 FAX: 707 363 690

Schwarz Pharma, Ltd.
Schwarz House, East Street
Chesham, Bucks HP5 1DG

Telephone: 494 772 071 FAX: 494 773 934

Searle
P.O. Box 53 Lane End Road
High Wycombe, Bucks HP12 4HL

Telephone: 494 521 124 FAX: 494 447 872

Smith & Nephew Pharmaceuticals, Ltd.
Bampton Road, Harold Hill
Romford, Essex RM3 8SL

Telephone: 708 349 333 FAX: 708 371 316

SmithKline Beecham plc
One New Horizons Court
Brentford, Middlesex TW8 9EP

Telephone: 81 975 2000 FAX: 81 847 0830

Key Personnel, This Location:
H. Wendt, Chairman
R.A. Clark, Vice Chairman

Stuart Pharmaceuticals
King's Court, Water Lane
Wilmslow, Cheshire SK9 5AZ

Telephone: 625 535 999 FAX: 625 712 450

Upjohn, Ltd.
Fleming Way
Crawley, West Sussex RH10 2LZ

Parent Company:
The Upjohn Company

Telephone: 293 531 133 FAX: 293 548 850

Wyeth Laboratories
Huntercombe Lane South, Taplow
Maidenhead, Berks SL6 0PH

Telephone: 628 604 377 FAX: 628 666 368

Zeneca Pharmaceuticals
Alderley House, Alderley Park
Macclesfield, Cheshire SK10 4TF

Telephone: 625 582 828 FAX: 625 585 022

---

49. UNITED STATES

A L Laboratories
One Executive Drive, Box 1399
Fort Lee, NJ 07024

Telephone: (201) 947 7444

Key Personnel, This Location:
James E. Flynn, Vice President
Mark W. Sykles, Vice President

A&S Pharmaceuticals Corp.
480 Barnum Avenue
Bridgeport, CT 06608

Telephone: (203) 368-2538

Key Personnel, This Location:
Arnold Lewis, President
Seth Lewis, Vice President

Abana Pharmaceuticals, Inc
1 Chase Corporate Dr., Suite 260
Birmingham, AL 35244

Telephone: (205) 988-4588. FAX: (205) 988-3294

Abbott Laboratories
1 Abbot Park Rd.
Abbott Park, IL 60064

Telephone: (800) 441-4987 FAX: (708) 937-1511

Key Personnel, This Location:
T.R. Hodgson President
P.N. Clark, Senior Vice President

Adams Laboratories, Inc.
14801 Soverign Road
Fort Worth, TX 76155

Telephone: (817) 545-7791 FAX: (817) 354-7820

Key Personnel, This Location:
John Q. Adams

Adria Laboratories
7001 Post Road
Dublin, OH 43017

Parent Company:
Procordia US, Inc.

Telephone: (614) 764-8100

Key Personnel, This Location:
Fernando DeCosta, President
Larry Versteegh, Exec. Vice President

Advanced Nutritional Technology, Inc.
1111 Jefferson Ave., P.O. Box 3225
Elizebeth, NJ 07207

Telephone: (908) 354-2740

Key Personnel, This Location:
Wayne L. Phillips

Advanced Polymer Systems, Inc.
3696 Haven Avenue
Redwood City, CA 94063

Telephone: (415) 366-2626 FAX: (415) 365-6490

Key Personnel, This Location:
Thomas J. Bigger, Sr. Vice President

Advanced Viral Research Corp.
300 Biscayne Boulevard Way
Miami, FL 33131

Telephone: (305) 347-4847

Key Personnel, This Location:
Bernard Friedland, President
William Bregman, Secretary & Treasurer

Advocare International, LLC
4100 Alpha, Suite 200
Dallas, TX 75244

Telephone: (214) 386-7782 FAX: (214) 386-6744

Key Personnel, This Location:
Ruth Ann Box

Akorn, Inc.
100 Akorn Drive
Abita Springs, LA 70420

Telephone: (800) 535-7155 FAX: (504) 893-1257

Key Personnel, This Location:
D.B. LeBlanc, President
E.M. Wingerter, Vice-President

Akzo America, Inc.
300 South Riverside Plaza
Chicago, IL 60606

Parent Company:
Akzo NV, Netherlands

Telephone: (312) 906-7500

Key Personnel, This Location:
Eugene Wilcaucus, President
Peter S. Gould, Vice President

Alcon Laboratories, Inc.
P.O. Box 6600
6201 South Freeway
Fort Worth, TX 76134

Parent Company:
Nestle SA

Telephone: (817) 293-0450

FAX: (817) 551-5485

Key Personnel, This Location:
Edgar H. Schollmaier, President
T.R.G. Sear, Exec. Vice President

Alkermes, Inc.
26 Landsdowne Street
Cambridge, MA 02139

Telephone: (617) 494-0171

Key Personnel, This Location:
Richard F. Pops, President
John W. Kozarich, Vice President

Allen & Hanburys
Five Moore Drive
Research Triangle Park, NC 27709

Parent Company:
Glaxo, Inc., USA

Telephone: (919) 248-2500

Allergan, Inc.
2525 Dupont Drive, P.O. Box 19534
Irvine, CA 92713

Telephone: (714) 752-4500 FAX: (714) 955-6987

Key Personnel, This Location:
W.C. Shepherd, President
R.M. Haugen, Exec. Vice President

Alliance Pharmaceutical Corp.
3040 Science Park Road
San Diego, CA 92121

Telephone: (619) 558-4300

Key Personnel, This Location:
Duane J. Roth, President
Theodore D. Roth, Exec. Vice President

Alpha 1 Biomedicals, Inc.
6903 Rockledge Drive
Bethesda, MD 20817

Telephone: (301) 564-4400

Key Personnel, This Location:
Vincent F. Simmon, President
Judith A. Hautala, Vice President

Alpha Therapeutic Corp.
5555 Valley Boulevard
Los Angeles, CA 90032

Telephone: (800) 421-0008 FAX: (213) 227-9053

Alra Laboratories, Inc.
3850 Clearview Court
Gurnee, IL 60031

Telephone: (708) 244-9440 FAX: (708) 244-9464

Altana, Inc.
60 Baylis Road
Melville, NY 11747

Parent Company:
Byk Gulden Lomberg GmbH

Telephone: (516) 454-7677

Key Personnel, This Location:
Rolf Rahmstorf, President
Arthur Dalik, Jr., Vice President

Alza Corp.
950 Page Mill Road
P.O. Box 10950
Palo Alto, CA 94303

Telephone: (800) 634-8977 FAX: (415) 494-5121

Key Personnel, This Location:
J.E. Shaw, President
A.M. Gerber, Exec. Vice President

American Home Products Corp.
Five Giralda Farms
Madison, NJ 07940

Telephone: (201) 660-5000 FAX: (201) 660-5771

Key Personnel, This Location:
John R. Stafford, Chairman & President
Stanley F Barshay, Sr. Vice President

American Pharmaceutical Co.
245 Fourth Street
Passaic, NJ 07055

Telephone: (201) 779-5300

Key Personnel, This Location:
Arthur Brown, President
Alfred Bagwell, Vice President

American Vitamin Products, Inc.
505 Halls Mill Road
Freehold, NJ 07728

Telephone: (908) 308-3000

Key Personnel, This Location:
E. Joseph Edell, President
I. Alan Hirschfield, Exec. Vice President

Amgen, Inc.
1840 DeHavilland Drive
Thousand Oaks, CA 91230

Telephone: (800) 772-6436 FAX: (800) 865-3707

Key Personnel, This Location:
K.W. Sharer, President
L.A. May, Vice President

Apothecon
P.O. Box 4500
Princeton, NJ 08543

Parent Company:
Bristol-Myers Squibb Company

Telephone: (609) 987-6800

Arco Pharmaceuticals, Inc.
105 Orville Drive
Bohemia, NY 11716

Telephone: (516) 567-9500

Parent Company:
Nature's Bounty, Inc.

Argus Pharmaceuticals, Inc.
3400 Research Forest Drive
Woodlands, TX 77381

Telephone: (713) 367-1666

Key Personnel, This Location:
David M. Leech, President
Kenneth M. Cohen, Vice President
Robert P. Lenk Vice President

Armour Pharmaceutical Company
500 Arcola Road, P.O. Box 1200
Collegeville, PA 19426

Telephone: (215) 454-8000

Parent Company:
Rhône-Poulenc, Inc.

FAX: (215) 454-8940

Armstromg Pharmaceuticals, Inc.
71 Elm Street
New Canaan, CT 06840

Telephone: (203) 966-4170

Key Personnel, This Location:
Harvey Mintzer, President
B.A. Shepherd, Exec. Vice President

B.F. Ascher & Company, Inc.
15501 W. 109th Street
Lexena, KS 66219

Telephone (913) 888-1880

Astra USA, Inc.
50 Otis Street
Westboro, MA 01581

Parent Company:
Astra AB

Telephone: (508) 366-1100 FAX: (508) 366-7406

Key Personnel, This Location:
Lars Bildman, President
Stefan Solvell, Treasurer

Athena Neurosciences, Inc.
800F Gateway Boulevard
South San Francisco, CA 94080

Telephone: (415) 877-0900 FAX: (415) 877-8370

Key Personnel, This Location:
J. Groom, President
P.E. Setler, Senior Vice President

Baker Cummins Dermatologicals
1950 Swarthmore Avenue
Lakewood, NJ 08701

Telephone: (908) 905-5200 FAX: (908) 905-7726

Baker Norton Pharmaceuticals, Inc.
8800 N.W. 36th Street
Miami, FL 33178

Telephone: (800) 374-4774 FAX: (305) 590-2252

Barr Laboratories
Two Quaker Road, P.O. Box 2900
Pomona, NY 10970

Telephone: (914) 362-1100

Key Personnel, This Location:
Bruce Downey, President
Gerald F. Price, Exec. Vice President

Barre-National, Inc.
7205 Windsor Boulevard
Baltimore, MD 21244

Parent Company:
A.L. Laboratories, Inc.

Telephone: (410) 298-1000

Key Personnel, This Location:
George S. Barrett, President
Thomas D. Darbey, Sr., Vice President

Barry Laboratories
2100 Park Central Boulevard, North, Suite 500
Pompano Beach, FL 33064

Telephone: (800) 327-1141

Basel Pharmaceuticals
556 Morris Ave.
Summit, NJ 07901

Parent Company:
Ciba-Geigy Corp.

Telephone: (908) 277-5000

Baxter International, Inc.
One Baxter Parkway
Deerfield, IL 60015

Telephone: (708) 948-2000

Key Personnel, This Location:
V..R. Loucks, Jr. Chairman & CEO
L.B. Knight, Exec. Vice President
T.L. White, Exec. Vice President

Beach Pharmaceuticals
5220 S. Manhattan Avenue
Tampa, Fl 33611

Parent Company:
Beach Products, Inc.

Telephone: (813) 839-6565

Key Personnel, This Location:
Richard S. Jenkins, Exec. Vice President
Victor De Oreo, Vice President, Sales

Beiersdorf, Inc.
360 Dr. Martin Luther King Drive
Norwalk, CT 06856

Parent Company:
Beiersdorf AG

Telephone: (203) 853-8008

FAX: (203) 854-8180

Key Personnel, This Location:
Peter M. Metzger, President
Charles Assicurato, Sr. Vice President
Chaucy Johnstone, Sr. Vice President

Belmac Corp.
4830 West Kennedy Boulevard
Tampa, FL 33609

Telephone: (813) 286-4401

Key Personnel, This Location:
Donald F. Boultace, President
Robert M. Stote, Vice President

Berlex Laboratories
300 Fairfield Road
Wayne, NJ 07470

Parent Company:
Schering Berlin, Inc.

Telephone: (201) 694-4100

Key Personnel, This Location:
Lutz Lingnau, Chairman
Wolfgang Kunze, Vice President

Berna Products Corp.
4216 Ponce de Leon Boulevard
Coral Gables, FL 33146

Telephone: (800) 533-5899

Key Personnel, This Location:
Andres Murai, Jr.

Beutlich L.P.
7149 N. Austin Avenue
Niles, IL 60714

Telephone: (800) 238-8542

FAX: (708) 647-8922

Biocraft Laboratories, Inc.
18-01 River Road, Box 948
Fair Lawn, NJ 07410

Telephone: (201) 703-0400

Key Personnel, This Location:
Harold Snyder, President
Beatrice Snyder, Exec. Vice President

Biogen, Inc.
14 Cambridge Center
Cambridge, MA 02142

Telephone: (617) 252-9200

Key Personnel, This Location:
James R. Tobin, President
Irving H. Fox, Vice President

Biopharmaceutics, Inc.
990 Station Road
Bellport, NY 11713

Telephone: (516) 286-5800

Key Personnel, This Location:
Edward Fine, President
William Kugler, Vice President

Blaine Company, Inc.
1465 Jamike Lane
Erlanger, KY 41018

Telephone: (606) 283-9437 FAX: (606) 283-9460

Key Personnel, This Location:
Alex M. Blaine

Block Drug Company, Inc.
257 Cornelison Avenue
Jersey City, NJ 07302

Telephone: (201) 434-3000

Key Personnel, This Location:
T.R. Block, President
T.H. LeSieur, Exec. Vice President

Bock Pharmacal Company
P.O. Box 419056
St. Louis, MO 63141

Telephone: (314) 579-0770

Boehringer Ingleheim Pharmaceuticals, Inc
900 Ridgebury Rd., P.O. Box 368
Ridgefield, CT 06877

Parent Company:
Boehringer Ingleheim AG,

Telephone: (203) 798-9988

Key Personnel, This Location:
Werner Gerstenberg, President
Holger Huels, Sr. Vice President

Boehringer Mannheim Pharmaceuticals
15204 Omega Drive
Rockville, MD 20850

Telephone: (301) 216-3750 FAX: (301) 330-7260

Boiron
1208 Amosland Road
Norwood, PA 19074

Telephone: (800) 258-8823

Key Personnel, This Location:
Ambroise Demonceaux, East Coast Manager

Bolar Pharmaceutical Company, Inc.
33 Ralph Avenue, Box 30
Copaigue, NY 11726

Telephone: (516) 842-8383

Key Personnel, This Location:
Lawrence S. Raisfeld, President
Jack J. Kornreich, Exec. Vice President

Boots Pharmaceuticals, Inc.
Suite 200
300 Tri-State International Center
Lincolnshire, IL 60069

Telephone: (708) 405-7400 FAX: (708) 405-7505

Key Personnel, This Location:
Carter H. Eckert, President
Neil M. Kuntz, Vice President

Bradley Pharmaceuticals, Inc.
383 Rte. 46 West
Fairfield, NJ 07004

Telephone: (201) 882-1505

Key Personnel, This Location:
Daniel Glassman, President
Iris Glassman, Treasurer

Braintree Laboratories, Inc.
P.O. Box 361
Braintree, MA 02184

Telephone: (617) 843-2202

Key Personnel, This Location:
Mr. Harry P. Keegan, President

Bristol Laborarories
P.O. Box 4500
Princeton, NJ 08543

Telephone: (609) 897-2000

Parent Company:
Bristol-Myers Squibb Company

Bristol-Myers Oncology Division
P.O. Box 4500
Princeton, NJ 08543

Telephone: (609) 892 2000

Parent Company:
Bristol-Myers Squibb Company,

Bristol-Myers Products
1350 Liberty Avenue
Hillside, NJ 07207

Parent Company:
Bristol-Myers Squibb Company

Telephone: (800) 468-7746

Bristol-Myers Squibb Company
345 Park Avenue
New York, NY 10154

Telephone: (212) 546-4000

Key Personnel, This Location:
C.A. Heimbold, Jr. President
M.E. Autera, Exec. Vice President

Burroughs Wellcome Company
3030 Cornwallis Road
Research Triangle Park, NC 27709

Parent Company:
The Wellcome Foundation, Ltd.

Telephone: (800) 722-9292

Key Personnel, This Location:
Philip R. Tracey, President
Gabriel R. Cipau, Vice President

Calcol, Inc.
301 West 57th Street
New York, NY 10019

Telephone: (212) 582-5861

Key Personnel, This Location:
N.C. Kaplan, President
D.P. Hochberg, Vice President

Calgon Vestal Laboratories, Inc.
7501 Page Avenue, P.O. Box 147
St. Louis, MO 63133

Parent Company:
Bristol-Myers Squibb Company

Telephone: (800) 325-8005

FAX: (314) 535-1064

Key Personnel, This Location:
Walter R. Maupay, Jr., President
T.J. Cini, Vice President

Cambridge Neuroscience, Inc.
One Kendall Square
Cambridge, MA 02139

Telephone: (617) 225-0600

Key Personnel, This Location:
Elkan R. Gamzu, President
Robert N. McBurney

Care-Tech Laboratories
3224 South Kingshighway Boulevard
St. Louis, MO 63139

Telephone: (314) 772-4610

FAX: (314) 772-4613

Key Personnel, This Location:
Sherry L. Brereton

J.R. Carlson Laboratories, Inc.
15 College Drive
Arlington Heights, IL 60004

Telephone: (708) 255 1600

FAX: (708) 255-1605

Carnick Laboratories, Inc.
65 Horse Hill Road
Cedar Knolls, NJ 07927

Telephone: (201) 267-2670 FAX: (201) 267-2728

Key Personnel, This Location:
Edmund Bergeron, President
Fred Weinstein, Vice President

Carrington Laboratories
2001 Walnut Hill Lane
Irving, TX 75038

Telephone: (214) 518-1300

Key Personnel, This Location:
Karl H. Meister, President
William H McAnalley, Vice President

Carter-Wallace, Inc.
1345 Avenue of the Americas
New York, NY 10105

Telephone: (212) 339-5000

Key Personnel, This Location:
Daniel J. Black, President
John Bridgen, Vice President

Center Laboratories
35 Channel Drive
Port Washington, NY 11050

Parent Company:
EM Industries, Inc., USA

Telephone: (800) 223-6837

Central Pharmaceuticals, Inc.
120 East Third Street
Seymour, IN 47274

Telephone: (812) 522-3915

Key Personnel, This Location:
Virginia B. Toulmin, President
Daniel J. Desmond, Exec. Vice President

Cerenex
5 Moore Drive
Research Triangle Park, NC 27709

Telephone: (919) 248-2100

Cetus Oncology Corp.
4560 Horton Street
Emeryville, CA 94608

Parent Company:
Chiron Corp.

Telephone: (800) 238-8779

FAX: (510) 601-3434

Cetylite Industries, Inc.
9051 River Road, Box 90006
Pennsauken, NJ 08110

Telephone: (609) 665-6111

FAX: (609) 665-5408

Key Personnel, This Location:
Stanley L. Wachman, President
Mollie Cohen, Secretary & Treasurer

Chantel Pharmaceutical Corp.
12400 Wilshire Boulevard
Los Angeles, CA 90025

Telephone: (310) 207-1950

Key Personnel, This Location:
Chantel Burnison, President
Rainer Stauback, Exec. Vice President

Chase Pharmaceutical Company
280 Chestnut Street
Newark, NJ 07105

Telephone: (201) 589-8181

Key Personnel, This Location:
Richard C. Remaly, President
Gregory Bowers, Exec. Vice President

Chattem, Inc.
1715 West 38th Street
Chattanooga, TN 37409

Telephone: (615) 821-4571

Key Personnel, This Location:
Z. Guerry, President
Robert E. Bosworth, Exec. Vice President

Chesebrough-Ponds USA
33 Benedict Place, P.O. Box 6000
Greenwich, CT 06836

Parent Company:
Unilever, Inc.

Telephone: (800) 243-5804

Key Personnel, This Location:
Richard G.F. Finn, President
Peter M. England, Exec. Vice President

Chiron Corp.
4560 Horton Street
Emeryville, CA 94608

Telephone: (510) 655-8730

Key Personnel, This Location:
Edward E. Peahoet, President
Renato Fuchs, Sr. Vice President

Ciba Consumer Pharmaceuticals
Mack Woodbridge II
581 Main Street
Woodbridge, NJ 07095

Parent Company:
Ciba-Geigy Corp.

Telephone: (908) 602-6000

FAX: (908) 602-6612

Key Personnel, This Location:
Douglas Brownstone

Ciba Pharmaceutical Company
556 Morris Avenue
Summit, NJ 07901

Parent Company:
Ciba-Geigy Corp.

Telephone: (908) 277-5000

Key Personnel, This Location:
Douglas G. Watson, President
Carolyn Koestenblatt, Vice President

Otis Clapp & Sons, Inc.
115 Shawmut Road, Box 9160
Canton, MA 02021

Telephone: (617) 821-5400

Key Personnel, This Location:
Donald J. Breen, President

Clint Pharmaceuticals, Inc.
1451 Elm Hill Pike, 252 Park Elm Complex
Nashville, TN 37210

Telephone: (800) 677-5022

Key Personnel, This Location:
Jeffery James

Colgate Oral Pharmaceuticals, Inc.
One Colgate Way
Canton, MA 02021

Parent Company:
Colgate-Palmolive Company

Telephone: (800) 225-3756

Key Personnel, This Location:
Ian Cook, Vice President

Colombia Laboratories, Inc.
2665 South Bayshore Drive
Miami, FL 33133

Telephone: (305) 860-1670

Connaught Laboratories, Inc.
Swiftwater, PA 18370

Telephone: (717) 839-4235

Parent Company:
Pasteur Merieux Company

Cortex Pharmaceuticals, Inc.
15241 Barranca Parkway
Irvine, CA 92718

Telephone: (714) 727-3171

Key Personnel, This Location:
Alan A. Steigrod, President
D.S. Hagen, Vice President

Copley Pharmaceutical, Inc.
25 John Road
Canton, MA 02021

Telephone: (617) 821-6111

Parent Company:
Hoechst Celanese Corp.

FAX: (617) 575-7471

Key Personnel, This Location:
Jane C.I. Hirsch, President
Bernard Grubstein, Exec. Vice President

Cumberland Manufacturing Company
13th & D Streets
Smyrna, TN 37167

Telephone: (615) 459-8900

Key Personnel, This Location:
Lonnie Smith, President
Randall Goldstein, Exec. Vice President

Parent Company:
Cumberland-Swan, Inc.

Curatek Pharmaceuticals
1965 Pratt Boulevard
Elk Grove Village, IL 60007

Telephone: (800) 332-7680

Cypros Pharmaceutical Corp.
11211 Sorrento Valley Road
San Diego, CA 92121

Telephone: (619) 455-6093

Key Personnel, This Location:
Paul A Morangos, President
Howard Jones, Vice President

Cytorad, Inc.
600 College Road East
Princeton, NJ 08540

Telephone: (609) 987-8270

Key Personnel, This Location:
M.D. Cleary, President

DDI Pharmaceuticals, Inc.
518 Logue Avenue
Mountain View, CA 94043

Telephone: (415) 964-7676

Key Personnel, This Location:
Raymond R. Rogers, President
Russell E Teasdale, Exec. Vice President

DTI Medical Corp.
310 East 4500 South
Salt Lake City, UT 84107

Telephone: (801) 263-3294

Key Personnel, This Location:
J. Randolf Harig, President
Frank E. Federer, Vice President

Daniels Pharmaceuticals, Inc.
2517 25th Avenue North
St. Petersburg, FL 33713

Telephone: (800) 237-7427

Dartmouth Pharmaceuticals, Inc.
37 Pine Hill Lane
Marion, MA 02738

Telephone: (508) 748-3209FAX: (508) 748-3132

Dayton Laboratories, Inc.
3307 Northwest 74th Avenue
Miami, FL 33122

Telephone: (305) 595-0988

Del Laboratories, Inc.
565 Broad Hollow Road
Farmingdale, NY 11735

Telephone: (516) 293-7070

Key Personnel, This Location:
D.K. Wassong, President
E.J. Vialardi, Sr. Vice President

Del Pharmaceuticals, Inc.
163 East Bethpage Road
Plainview, NY 11803

Telephone: (516) 293-7070

Key Personnel, This Location:
Peter Liman, Vice President

Parent Company:
Del Laboratories, Inc.

FAX: (516) 293-9018

Denison Pharmaceuticals, Inc.
60 Dunnell Lane
Pawtucket, RI 02860

Telephone: (401) 723-5500

Key Personnel, This Location:
Edward S. Stone, President
Barbara Stone, Vice President

Dermik Laboratories, Inc.
500 Arcola Road, P.O. Box 1200
Collegeville, PA 19426

Telephone: (610) 454-8000

Dey Laboratories, Inc.
2751 Napa Valley Corporate Drive
Napa, CA 94558

Telephone: (800) 755-5560 FAX: (707) 224-3235

Key Personnel, This Location:
Susan Snars

Dista Products Company
Lilly Corporate Center
Indianapolis, IN 46285

Parent Company:
Eli Lilly & Company

Telephone: (317) 276-4000

Doak Pharmacal Company, Inc.
67 Sylverter Street
Westbury, NY 11590

Telephone: (516) 333-7722

Key Personnel, This Location:
Lorna Adams, President

Dunhill Pharmaceuticals, Inc.
P.O. Box 100
Gavette, AR 72736

Telephone: (501) 787-5232

Key Personnel, This Location:
M.J. Staha, President
J.H. Hatfield, Vice President

DuPont Merck Pharmaceutical Company
DuPont Merck Plaza
Routes 48 & 141
Wilmington, DE 19880

Telephone: (302) 992-5107 FAX: (302) 992-7771

Key Personnel, This Location:
Joseph A. Mollica, CEO
Richard J. Davis, Vice President

Dura Pharmaceuticals, Inc.
11175 Flintkote Avenue
San Diego, CA 92121

Telephone: (619) 457-2553

Key Personnel, This Location:
C.L. Garner, President
W. F. Sparth, Vice President

Duramed Pharmaceuticals, Inc.
5040 Lester Road
Cincinnati, OH 45213

Telephone: (513) 731-9900

Key Personnel, This Location:
E. Thomas Arington, President
Doane Darling, Vice President

ECR Pharmaceuticals
3911 Deep Rock Road, P.O. Box 71600
Richmond, VA 23255

Telephone: (804) 527-1950 FAX: (804) 527-1959

Ebamont NV
P.O. Box 16529
Colombus, OH 43216

Telephone: (614) 764-8100

Key Personnel, This Location:
Fernanda DeCosta, President
Romeo Roncucci, Vice President

Effcon Laboratories, Inc.
P.O. Box 7509
Marietta, GA 30065

Telephone: (800) 722-2428 FAX: (404) 499-0058

Key Personnel, This Location:
Leigh Ann Buice

Elder Pharmaceuticals, Inc.
3300 Hyland Avenue
Costa Mesa, CA 92626

Telephone: (714) 545-0100

Key Personnel, This Location:
Adam Jerney, President

Parent Company:
SPI Pharmaceuticals, Inc.

Elkins-Sinn, Inc.
2 Esterbrook Lane
Cherry Hill, NJ 08003

Telephone: (609) 434-3700

Key Personnel, This Location:
L.J. Hayko, President
W. Hartnett, Vice President

Parent Company:
A.H. Robins Company, Inc.

Enzon, Inc.
40 Kingsbridge Road
Piscataway, NJ 08854

Telephone: (908) 980-4634
FAX: (908) 980-5911

Key Personnel, This Location:
Abraham Abuchowski, President
E. Quinton Gordon, Vice President

Everett Laboratories, Inc.
71 Glenwood Place
East Orange, NJ 07017

Telephone: (210) 674-8455
FAX: (201) 674-0933

Evergood Products Corp.
175 Lauman Lane
Hicksville, NY 11801

Telephone: (516) 822-1230

Key Personnel, This Location:
Sidney Rich, President
Melvin L. Rich, Exec. Vice President

Ferndale Laboratories, Inc.
780 West Eight Mile Road
Ferndale, MI 48220

Telephone: (313) 548-0900 FAX: (313) 548-0708

Key Personnel, This Location:
Thayer McMillan

Ferring Laboratories, Inc.
400 Rella Boulevard, Suite 201
Suffern, NY 10901

Telephone: (914) 368-7900

The Fielding Pharmaceutical Company, Inc.
94 Weldon Parkway, P.O. Box 2186
Maryland Heights, MO 63043

Telephone (314) 567-5462

Fisons Corporation
755 Jefferson Road, P.O. Box 1766
Rochester, NY 14623

Parent Company:
Fisons plc

Telephone: (716) 475-9000

C.B. Fleet & Co., Inc.
4615 Murray Place
Lynchburg, VA 24502

Telephone: (804) 528-4000

Key Personnel, This Location:
Brian Duffy, President

Fleming & Company
1600 Fenpark Drive
Fenton, MO 63026

Telephone: (314) 343-8200

Key Personnel, This Location:
John J. Roth

Forest Pharmaceuticals, Inc.
2510 Metro Boulevard
St. Louis, MO 63043

Telephone: (314) 569-3610

Parent Company:
Forest Laboratories, Inc.

Fountain Pharmaceuticals, Inc.
11201 Concept Boulevard
Largo, FL 34643

Telephone: (813) 392-5300

Key Personnel, This Location:
John C. Walsh, President

Fujisawa USA, Inc.
Three Parkway North Center
Deerfield, IL 60015

Parent Company:
Fujisawa Pharmaceutical Co., Ltd.

Telephone: (800) 727-7003

Key Personnel, This Location:
Takashi Aoki, Chairman
R. Richard Wieland, Sr. Vice President

G & W Laboratories, Inc.
111 Coolidge Street
South Plainfield, NJ

Telephone: (908) 753-2000

Key Personnel, This Location:
Burton G. Greenblat, President
A.N. Rao, Exec. Vice President

Galderma Laboratories, Inc.
P.O. Box 331329
Fort Worth, TX 76163

Telephone: (817) 263-2600

Gate Pharmaceuticals
650 Cathill Road, P.O. Box 904
Sellersville, PA 18960

Parent Company:
Lemmon Company

Telephone: (215) 256-8400

FAX: (215) 256-7855

Gebauer Company
9410 St. Catherine Avenue
Cleveland, OH 44104

Telephone: (800) 321-9348   FAX: (216) 271-5335

Key Personnel, This Location:
John A. Giltiman, President

Geigy Pharmaceuticals
556 Morris Avenue
Summit, NJ 07901

Parent Company:
Ciba-Geigy Corp.

Telephone: (908) 277-5000

Genderm Corp.
600 Knightsbridge Parkway
Lincolnshire, IL 60069

Telephone: (708) 634-7373

Genentech, Inc.
460 Point San Bruno Boulevard
South San Francisco, CA 94080

Parent Company:
Roche Holdings, Inc.

Telephone: (415) 225-1000

Key Personnel, This Location:
G. Kirk Raab, President
William D. Young, Exec. Vice President

Geneva Pharmaceuticals, Inc.
2555 West Midway Boulevard, P.O. Box 446
Broomfield, CO 80038

Telephone: (303) 466-2400 FAX: (303) 469-6467

Key Personnel, This Location:
Charles Lay, President
Joseph Papa, Sr. Vice President

Gensia, Inc.
11025 Roselle Street
San Diego, CA 92121

Telephone: (619) 456-8300

Key Personnel, This Location:
D.F. Hale, President
Carl Bobkoski, Exec. Vice President

Genta, Inc.
3550 General Atomics Court
San Diego, CA 92121

Telephone: (619) 455-2700

Key Personnel, This Location:
W.E. Bliss, President
Donald Picker, Exec. Vice President

Genzyme Corp.
One Kendall Square
Cambridge, MA 02139

Telephone: (800) 745-4447 FAX: (617) 252-7600

Key Personnel, This Location:
Henri A. Termeer, President
Joseph Papa, Sr. Vice President
Geoffrey F. Cox, Sr. Vice President

Glaxo Dermatology
Five Moore Drive
Research Triangle Park, NC 27709

Parent Company:
Glaxo, Inc.

Telephone: (919) 248-2100

Glaxo Pharmaceuticals
Five Moore Drive
Research Triangle Park, NC 27709

Parent Company:
Glaxo, Inc.

Telephone: (919) 248-2100

Glenbrook Laboratories
90 Park Avenue
New York, NY 10016

Parent Company:
Sterling Winthrop, Inc,

Telephone: (212) 907-2000

Key Personnel, This Location:
Robin D. Mills, President
J.T. Whitmer, Vice President

Glenwood, Inc.
83 North Summit Street
Tenafly, NJ 0767

Telephone: (201) 569-0050

Gordon Laboratories
6801 Ludlow Street
Upper Darby, PA 19082

Telephone: (215) 734-2011 FAX: (215) 734-2049

Key Personnel, This Location:
David Dercher

Gray Pharmaceutical Co.
100 Connecticut Avenue
Norwalk, CT 06850

Parent Company:
The Purdue Fredrick Company

Telephone: (203) 853-0123

Greer Laboratories, Inc.
639 Nuway Circle
Lenoir, NC 28645

Telephone: (704) 754-5327

Key Personnel, This Location:
William White, Jr., President
R.E. Esch, Vice President

Guardian Laboratories
230 Marcus Boulevard
Hauppauge, NY 11788

Parent Company:
United-Guardian, Inc.

Telephone: (800) 645-5566

Gynex Pharmaceuticals, Inc.
1175 Corporate Woods Parkway
Vernon Hills, IL 60061

Telephone: (708) 913-1144

Key Personnel, This Location:
S.M. Simes, President
B.N. Barron, Vice President

Parent Company:
Biotechnology General Corp.

Gynopharma, Inc.
50 Division Street
Somerville, NJ 08876

Telephone: (908) 725-3100

Halsey Drug Company
1827 Pacific Street
Brooklyn, NY 11233

Telephone: (718) 467-7500

Key Personnel, This Location:
R. Foran, President

Health Maintainence Programs, Inc.
7 Westchester Plaza
Elmsford, NY 10523

Telephone: (800) 362-8673

Health Products International Corp.
P.O. Box 4000
Springville, UT 84663

Parent Company:
Murdock Holding Company

Telephone: (801) 377-3093

Key Personnel, This Location:
Derek Hall, President
Hugo Boren, Vice President

Herald Pharmacal, Inc.
6503 Warwick Road
Richmond, VA 23225

Telephone: (804) 745-3400

FAX: (804) 745-1963

Key Personnel, This Location:
T.A. Bakewell, President
Gary Palmer, Marketing Director

Dow B. Hickam, Inc.
P.O. Box 2006
Sugar Land, TX 77487

Telephone: (713) 240-1000

High Chemical Company
3901-A Nebraska Street
Levittown, PA 19056

Telephone: (800) 447-8792

Hill Dermaceuticals, Inc.
505 West Robinson Street
Orlando, FL 32801

Telephone (407) 896-8280 FAX: (407) 649-9213

Key Personnel, This Location:
Rosario G. Ramirez

Hi-Tech Pharmacal
369 Bayview Avenue
Amityville, NY 11701

Telephone: (516) 789-8228 FAX: (516) 789-8429

Key Personnel, This Location:
Bernard Seltzer, President
David S. Seltzer, Exec. Vice President

Hoechst-Roussel Pharmaceuticals, Inc.
Route 202-206, P.O. Box 2500
Somerville, NJ 08760

Telephone: (800) 451-4455

Hoffmann-LaRoche, Inc.
340 Kingsland Street
Nutely, NJ 07110

Parent Company:
Roche Holdings, Inc.

Telephone: (201) 235-5000

Key Personnel, This Location:
P.J. Zenner, President
J. Drews, Sr. Vice President
R.E. Norton, Vice President

Horus Therapeutics, Inc.
2320 Brighton-Henrietta Town Line Road
Rochester, NY 14623

Telephone: (716) 292-4820 FAX: (716) 292 4836

ICN Pharmaceuticals, Inc.
ICN Plaza, 3300 Hyland Avenue
Costa Mesa, CA 92626

Telephone: (800) 556-1937

Key Personnel, This Location:
Milan Panic, Chairman
Kobi Sethna, Vice President
John Julian, Vice President

ICOS Corp.
22021 20th Avenue Southeast
Bothell, WA 98021

Telephone: (206) 485-1900

Key Personnel, This Location:
George B. Rathmann, President
C. S. Henney, Exec. Vice President

Immunex Corp.
51 University Street
Seattle, WA 98101

Telephone: (800) 466-8639 FAX: (206) 343-8926

Key Personnel, This Location:
M. Kranda, President
S. Gillis, Exec. Vice President

Immuno-U.S., Inc.
1200 Parkdale Road
Rochester, MI 48307

Telephone: (313) 652-7872 FAX: (313) 652-0670

Inter-Cal Corp.
533 Madison Avenue
Prescott, AZ 86301

Telephone: (602) 445-8063

Key Personnel, This Location:
Gerald W. Elders

Ion Laboratories, Inc.
7431 Pebble Drive
Fort Worth, TX 76118

Telephone: (817) 589-7257 FAX: (817) 590-0973

Key Personnel, This Location:
Ralph E. Brown

Ivax Corp.
8800 Northwest 36th Street
Miami, FL 33178

Telephone: (305) 590-2200

Key Personnel, This Location:
Phillip Frost, President
F.A. Baxter, Sr. Vice President

Jacobus Pharmaceutical Co., Inc.
37 Cleveland Lane, P.O. Box 5290
Princeton, NJ 08540

Telephone: (609) 921-7447

Jamol Laboratories, Inc.
13 Ackerman Avenue
Emerson, NJ 07630

Telephone: (201) 262-6363

Janssen Pharmaceutica, Inc.
1125 Trenton-Harbourton Road, P.O. Box 200
Titusville, NJ 08560

Telephone: (609) 730-2000 FAX: (609) 730-3044

Key Personnel, This Location:
David C. Roche, Vice President
Douglas N. Dobak, Vice President
Henry Avallone, Ex. Director

Johnson & Johnson
Consumer Products, Inc.
Grandview Road
Skillman, NJ 08558

Parent Company:
Johnson & Johnson

Telephone: (800) 526-3967

Johnson & Johnson Medical, Inc.
P.O. Box 90130
Arlington, TX 76004

Parent Company:
Johnson&Johnson

Telephone: (800) 433-5009

Johnson & Johnson • Merck Consumer Pharmaceuticals Co.
Camp Hill Road
Fort Washington, PA 19034

Telephone: (215) 233-7000

Jones Medical Industries, Inc.
11601 Lilburn Park Road
St. Louis, MO 63146

Telephone: (314) 432-3785

Key Personnel, This Location:
Dennis M. Jones, President
Judith A. Jones, Secretary & Treasurer

Kabi Pharmacia
800 Centennial Avenue
P.O. Box 1327
Piscataway, NJ 08855

Parent Company:
Procordia US, Inc.

Telephone: (908) 457-8000

FAX: (908) 457-8283

Key Personnel, This Location:
M. Myslinski, President

Kenwood Laboratories
383 Route 46 West
Fairfield, NJ 07004

Parent Company:
Bradley Pharmaceuticals, Inc.

Telephone: (201) 882-1505

Key Pharmaceuticals, Inc.
Galloping Hill Road
Kenilworth, NJ 07033

Telephone: (800) 526-4099

Knight Industries, Inc.
750 East Sample Road
Pompano Beach, FL

Telephone: (305) 942-8708

Key Personnel, This Location:
William McMurray, President
R.F. Phillips, Vice President

Knoll Pharmaceutical Company
30 North Jefferson Road
Whippany, NJ 07981

Parent Company:
BASF AG

Telephone: (201) 887-8300

Key Personnel, This Location:
G.E. Bendela, President
Fritz Frickel, Vice President

Konsyl Pharmaceuticals, Inc.
400 South Hulen
Fort Worth, TX 76109

Telephone: (817) 763-8011
FAX: (817) 731-9389

Key Personnel, This Location:
William Steiber

Kramer Laboratories, Inc.
8778 Southwest 8th Street
Miami, FL 33174

Telephone: (800) 824-4894

Lannett Company, Inc.
9000 State Road
Philadelphia, PA 19136

Telephone: (215) 333-9000

Key Personnel, This Location:
Barry Weisberg, President

Laser, Inc.
2200 West 97th Place, P.O. Box 905
Crown Point, IN 46307

Telephone: (219) 663-1165

Key Personnel, This Location:
Donald A. Laser

Lederle Consumer Health
One Cyanamid Plaza
Wayne, NJ 07470

Telephone: (800) 282-8805

Parent Company:
American Home Products Co.

Lederle Laboratories
One Cyanamid Plaza
Wayne, NJ 07470

Telephone: (800) 820-2815

Parent Company:
American Home Products Co.

Lemmon Company
650 Cathill Road, P.O. Box 630
Sellersville, PA 18960

Parent Company:
Teva Pharmaceuticals

Telephone: (800) 545-8000

Key Personnel, This Location:
William A. Fletcher, President
Ronald L. Nedich, Sr. Vice President

Eli Lilly & Company
Lilly Corporate Center
Indianapolis, IN 46285

Telephone: (317) 276-2000

Key Personnel, This Location:
Randall L. Tobais, Chairman & CEO
Sidney Taurel, Exec. Vice President

Lotus Biochemical Corp.
Kirk Bldg, Main Street, P.O. Box 126
Bland, VA 24315

Telephone: (703) 688-4711
FAX: (703) 688-4178

Key Personnel, This Location:
Jay Kirk

Lunsco, Inc.
Route 2, Box 42
Pulaski, VA 24301

Telephone: (703) 980-4358

MGI Pharma, Inc.
Opus Center, Suite 300-E
9900 Bren Road East
Minneapolis, MN 55343

Telephone: (800) 562-5580 FAX: (612) 935-0468

3M Pharmaceuticals
3M Center 275-3W-01, P.O. Box 33275
St. Paul, MN 55144

Telephone: (800) 328-0255

Key Personnel, This Location:
W.G. Meredith, Group Vice President

Magainin Pharmaceuticals, Inc.
5110 Campus Drive
Plymouth Meeting, PA 19462

Telephone: (215) 941-4020

Key Personnel, This Location:
Jay Moorin, President
L.S. Jacob, Exec. Vice President

Marion Merrell Dow, Inc.
9300 Ward Parkway, P.O.Box 8480
Kansas City, MO 64114

Parent Company:
Dow Chemical Company

Telephone: (800) 552-3656

Key Personnel, This Location:
Richard J Markham, President
Frank L. Douglas, Exec. Vice President
Francois Nader, Vice President

Marlyn Health Care
14851 North Scottsdale Road
Scottsdale, AZ 85254

Telephone: (602) 991-0200

Mayor Pharmaceutical Laboratories
2401 South 24th Street
Phoenix, AZ 85034

Telephone: (602) 244-8899

Mayrand Pharmaceuticals, Inc.
4 Dundas Circle, P.O. Box 8869
Greensboro, NC 27419

Telephone: (800) 334-0514

McNeil Consumer Products Company
Camp Hill Road
Fort Washington, PA 19034

Telephone: (215) 233-7000

Key Personnel, This Location:
James T. Lenehan, President
Patricia Stewart, Vice President

Parent Company:
Johnson & Johnson

McNeil Pharmaceutical
1000 Route 202
Raritan, NJ 08869

Telephone: (908) 218-6000

Parent Company:
Johnson & Johnson

Mead Johnson Nutritional Group
2400 West Lloyd Expressway
Evansville, IL 47721

Telephone: (812) 429-5000

Key Personnel, This Locsation:
E.L. Johnson, President
W.G McCabe, Vice President

Parent Company:
Bristol-Myers Squibb Company

Medicis Pharmaceutical Corp.
4343 East Camelback Road
Phoenix, AZ 85018

Telephone: (602) 808-8800

Key Personnel, This Location:
Harold E. Chappelear, President
Leonard L. Mazur, Exec. Vice President

Medtronic, Inc., Neurological Division
800 53rd Avenue, Northeast
Minneapolis, MN 55421

Telephone: (612) 572-5000

Menley & James, Inc.
100 Tournament Drive
Horsham, PA 19044

Telephone: (215) 441-6500

Key Personnel, This Location:
Lawrence D. White, President

Merck & Company
One Merck Drive, Box 100
Whitehouse Station, NJ 08889

Telephone: (908) 423-1000

Key Personnel, This Location:
R.V. Gilmartin, Chairman, President & CEO
J.T. Jackson, Exec. Vice President

Miles, Inc.,
Consumer HealthCare Products
1127 Myrtle Street
Elkhart, IN 46515

Parent Company:
Bayer AG

Telephone: (800) 800-4793

Key Personnel, This Location:
Gary S. Balkema, President
Horst K.D. Wallrabe, Exec. Vice President

Miles, Inc., Pharmaceutical Division
400 Morgan Lane
West Haven, CT 06516

Telephone: (203) 937-2000

Milex Products, Inc.
5915 Northwest Expressway
Chicago, IL 60631

Telephone: (312) 631-6484

Mission Pharmacal Company
1325 East Durango
San Antonio, TX 78210

Telephone: (512) 533-7118

Muro Pharmaceutical, Inc.
890 East Street
Tewksbury, MA 01876

Telephone: (508) 851-5981

Key Personnel, This Location:
George D. Behrakis, President

Mylan Pharmaceuticals, Inc.
781 Chestnut Ridge Road, P.O. Box 4310
Morgantown, WV 26504

Telephone: (304) 599-2595

Key Personnel, This Location:
Milan Puscar, President
Dana G. Barnett, Exec. Vice President

Natren, Inc.
3105 Willow Lane
Westlake Village, CA 91361

Telephone: (800) 992-3323 FAX: (805) 371-4742

Nature's Bounty, Inc.
90 Orville Drive
Bohemia, NY 11716

Telephone: (516) 567-9500 FAX: (516) 563-7148

Key Personnel, This Location:
S. Rudolf, Chairman & President
H. Kamil, Exec. Vice President

Neurogenesis, Inc.
11511 Katy Freeway
Houston, TX 77079

Telephone: (713) 496-0500

Key Personnel, This Location:
Thomas Chnielewski, President

Neutrogena Corp.
5760 West 96th Street
Los Angeles, CA 90045

Parent Company:
Johnson & Johnson

Telephone: (310) 642-1150 FAX: (310) 337-5557

Key Personnel, This Location:
Allan Kurtzman, President

Niche Pharmaceuticals, Inc.
300 Trophy Club Drive, # 400
Roanoke, TX 76262

Telephone: (817) 491-2770 FAX: (817) 491-3533

Key Personnel, This Location:
Steve F. Brandon

Northhampton Medical, Inc.
3039 Amwiler Road, Suite 122
Atlanta, GA 30360

Telephone: (404) 416-8889 FAX: (404) 416-0633

Novo-Nordisk Pharmaceuticals, Inc.
100 Overlook Center, Suite 200
Princeton, NJ 08540

Parent Company:
Novo-Nordisk A/S

Telephone: (609) 987-5800

Key Personnel, This Location:
C. Henk Bleeker, President

Novopharm, USA, Inc.
165 East Commerce Drive, Suite 200
Schaumburg, IL 60173

Telephone: (800) 635-5067 FAX: (708) 882-4232

Key Personnel, This Location:
Robert J. Gunther, President

Nutripharm Laboratories, Inc.
8 Bartles Corner Road, Suite 101
Flemington, NJ 08822

Telephone: (908) 806-8954 FAX: (908) 806-8934

Key Personnel, This Location:
S. Rao Kolli

Oclassen Pharmaceuticals, Inc.
100 Pelican Way
San Rafael, CA 94901

Telephone: (800) 288-4508

Ohm Laboratories, Inc.
P.O. Box 279
Franklin Park, NJ 08823

Telephone: (908) 297-3030

Key Personnel, This Location:
Arun Heble

Organon, Inc.
375 Mount Pleasant Avenue
West Orange, NJ 07052

Telephone: (201) 325-4500

Ortho Pharmaceutical Corp.
Route 202, P.O. Box 300
Raritan, NJ 08869

Telephone: (800) 682-6532

PDK Laboratories, Inc.
145 Ricefield Lane
Happauge, NY 11788

Telephone: (516) 273-2630

Key Personnel, This Location:
M.B. Krasnoff, President

PD-Rx Pharmaceuticals, Inc.
6000 Northwest Second Street
Oklahoma City, OK 73127

Telephone: (405) 495-3545

Key Personnel, This Location:
R.D. Halsey, President
R.R. Tutor, Vice President

PRN Laboratories, Inc.
1275 Bennett Drive, Suite 122
Longwood, FL 32750

Telephone: (407) 260-2225 FAX: (800) 761-1210

Paddock Laboratories, Inc.
3101 Louisiana Avenue North
Minneapolis, MN 55427

Telephone: (800) 328-5113

Palisades Pharmaceuticals, Inc.
64 North Summit Street
Tenafly, NJ 07670

Telephone: (201) 569-8502

Parke-Davis
201 Tabor Road
Morris Plains, NJ 07950

Telephone: (201) 540-2000

Parent Company:
Warner-Lambert Company

Pedinol Pharmacal, Inc.
30 Banfi Plaza North
Farmingdale, NY 11735

Telephone: (516) 293-9500

Penederm, Inc.
320 Lakeside Drive, Suite A
Foster City, CA 94404

Telephone: (415) 358-0100

Key Personnel, This Location:
B. Chaudhuri, Technical Director
T. Gutshall, Director of Operations

Penex Laboratories
One Penex Drive
Verona, PA 15147

Telephone: (412) 828-2900

Parent Company:
Rexall-Sundown, Inc.

FAX: (412) 826-4723

Key Personnel, This Location:
Elliot Prillig, President

Perrigo Company
117 Water Street
Allegan, MI 49010

Telephone:

Key Personnel, This Location:
Michael Jandernoa, President
Ralph Klingenmeyer, Exec. Vice President

Person & Covey, Inc.
616 Allen Avenue, P.O. Box 25018
Glendale, CA 91221

Telephone: (800) 423-2341

Key Personnel, This Location:
Lorne V Person, President

Pfizer Consumer Health Care
100 Jefferson Road
Parsippany, NJ 07054

Parent Company:
Pfizer, Inc.

Telephone: (201) 887-2100

Key Personnel, This Location:
B. J. Robinson, President

Pharmaceutical Resources, Inc.
One Ram Ridge Road
Spring Valley, NY 10977

Telephone: (914) 425-7100

Key Personnel, This Location:
Kenneth I. Sawyer, President
E.G. Shami, Sr. Vice President

Pharma-Singer, Inc.
23201 Via Celeste
Coto de Caza, CA 92679

Telephone: (714) 858-5250 FAX: (714) 858-3095

Pharmavite Corp.
15451 San Fernando Mission Boulevard, P.O. Box 9606
Mission Hills, CA 91346

Telephone: (800) 423-2405

PolyMedica Pharmaceuticals (USA), Inc.
2 Constitution Way
Woburn, MA 01801

Parent Company:
PolyMedica Indistries, Inc.

Telephone: (617) 933-2020

FAX: (617) 933-7992

Key Personnel, This Location:
Arthur Sicilliano

Porton Products Limited
30401 Agoura Road # 102
Agoura Hills, CA 91301

Telephone: (818) 879-2200

Pratt Pharmaceuticals
235 East 42nd Street
New York, NY 10017

Parent Company:
Pfizer, Inc.

Telephone: (212) 573-2422

Premier, Inc.
Greenwich Office Park One
Greenwich, CT 06831

Telephone: (203) 622-1211

FAX: (203) 622-0773

Key Personnel, This Location:
Robert Albus

Princeton Pharmaceutical Products
P.O. BOX 4500
Princeton, NJ 08543

Telephone: (609) 897-2000

Parent Company:
Bristol-Myers Squibb Company

Procter & Gamble Company
One Procter & Gamble Plaza
Cincinnati, OH 45202

Telephone: (513) 983-1000

Key Personnel, This Location:
Thomas A. Moore, President, Health Care Products
Mark A. Collar, Vice President & General Manager

Procter & Gamble Pharmaceuticals, Inc.
17 Eaton Avenue
Norwich, NY 13815

Telephone: (800) 448-4878

Parent Company:
Procter & Gamble Company

Puerto Rico Alcon
P.O. Box 3000
Humacao, Puerto Rico 00661

Parent Company:
Alcon Laboratories, Inc. USA

The Purdue Fredrick Company
100 Connecticut Avenue
Norwalk, CT 06850

Telephone: (203) 853-0123

Purepac Pharmaceuticals Company
200 Elmora Avenue
Elizabeth, NJ 07202

Parent Company:
Moleculon, Inc.

Telephone: (908) 527-9100

Key Personnel, This Location:
M.R.D. Ashton, President

Quad Pharmaceuticals, Inc.
6340 LaPass Trail
Indianapolis, IN 46268

Telephone: (317) 299-6611

Key Personnel, This Location:
K. Sawyer, President
L. Hoffman, Sr. Vice President

R&D Laboratories, Inc.
4094 Glencoe Avenue
Marina del Rey, CA 90292

Telephone: (310) 305-8053

Key Personnel, This Location:
Rhoda Markoff

Reckitt & Colman Pharmaceuticals, Inc.
1901 Huguenot Road
Richmond, VA 23235

Parent Company:
Reckitt & Colman Ltd.

Telephone: (804) 379-1090

FAX: (804) 379-1215

Reed & Camrick
257 Cornelison Avenue
Jersey City, NJ 07302

Parent Company:
Block Drug Company, Inc.

Telephone: (201) 434-4000
FAX: (201) 434-3032

Key Personnel, This Location:
John T. Spitznagel, President
P. Volk, Vice President

The Reese Chemical Company
10617 Frank Avenue
Cleveland, OH 44106

Telephone: (216) 231-6441
FAX: (216) 231-6444

Key Personnel, This Location:
George W. Reese, III, President
Sandra J. Reese, Vice President

Requa, Inc.
1 Seneca Place, P.O. Box 4008
Greenwich, CT 06830

Telephone: (203) 869-2445
FAX: (203) 661-5630

Key Personnel, This Location:
J. Geils

Research Industries Corp., Pharmaceutical Division
6864 South 300 West
Midvale, UT 84047

Telephone: (801) 562-0200
FAX: (801) 562-1122

Key Personnel, This Location:
R.D. Hibbert

Rexar Pharmacal Corp.
396 Rockaway Avenue
Valley Stream, NY 11581

Telephone: (516) 473-0222

Rhône-Poulenc Rorer Pharmaceuticals, Inc.
500 Arcola Road
Collegeville, PA 19426

Parent Company:
Rhône-Poulenc, SA

Telephone: (215) 454-8000

Key Personnel, This Location:
Michel deRosen, President

Richwood Pharmaceutical Company, Inc.
P.O. Box 6497
Florence, KY 41022

Telephone: (606) 282-2100 FAX: (606) 282-2103

Roberts Pharmaceutical Corp.
4 Industrial Way West
Eatontown, NJ 07724

Telephone: (908) 389-1182 FAX: (908) 389-1014

Key Personnel, This Location:
Robert A Vukovich, President
Louis C. Wines, Vice President

A.H. Robins Consumer Products Division
1407 Cummings Drive
Richmond, VA 23220

Telephone: (804) 257-2000

Parent Company:
American Home Products Corp.

Ross Products Division
Colombus, OH 43215

Telephone: (614) 624-7677

Key Personnel, This Location:
H.S. Sauls, Vice President

Parent Company:
Abbott Laboratories

Roxanne Laboratories, Inc.
1809 Wilson Road
Colombus, OH 43228

Telephone: (614) 276-4000

Rydelle Laboratories
1525 Howe Street
Racine, WI 53403

Telephone: (414) 631-4709

Key Personnel, This Location:
Marilyn K. Blood

Parent Company:
S.C. Johnson & Son, Inc.

FAX: (414) 631-3829

Sandoz Pharmaceuticals/ Consumer Division
59 Route 10
East Hanover, NJ 07936

Parent Company:
Sandoz AG

Telephone: (201) 503-7500

FAX: (201) 503-8256

Key Personnel, This Location:
T. G. Rothwell, President
Robert G. Donovan, Sr. Vice President

Sanofi Winthrop Pharmaceuticals
90 Park Avenue
New York, NY 10016

Telephone: (212) 907-2000

Savage Laboratories
60 Baylis Road
Melville, NY 11747

Parent Company:
Altana, Inc.

Telephone: (800) 231-0206

FAX: (516) 454-0732

Scandipharm, Inc.
22 Inverness Center Parkway, Suite 310
Birmingham, AL 35242

Telephone: (205) 991-8085

FAX: (205) 991-9547

Key Personnel, This Location:
D. Brakhage, Manager

Schein Pharmaceutical, Inc.
100 Campus Drive
Florham Park, NJ 07932

Telephone: (800) 537-2980 FAX: (516) 686-3010

Schering Corp.
Galloping Hill Road
Kenilworth, NJ 07033

Telephone: (908) 298-4000

Parent Company:
Schering-Plough Corp.

Schering-Plough Healthcare Products
110 Allen Road
Liberty Corner, NJ 07938

Telephone: (908) 604-1969

Parent Company:
Schering-Plough Corp.

Key Personnel, This Location:
David E. Collins, President

Scot-Tussin Pharmacal Company, Inc.
50 Clemence Street
Cranston, RI 02920

Telephone: (401) 942-8555 FAX: (401) 942-5690

Schwarz Pharma
5600 West County Line Road
P.O. Box 2038
Milwaukee, WI 53201

Telephone (414) 354-4300

Parent Company:
Kremers Urban Compny

G.D Searle & Company
P.O. Box 5110
Chicago, IL 60680

Telephone: (800) 323-1603

Serono Laroratories, Inc.
100 Longwater Circle
Norwell, MA 02061

Telephone: (617) 982-9000

Shaman Pharmaceuticals, Inc.
213 East Grand Avenue
South San Francisco, CA 94080

Telephone: (415) 952-7070

Key Personnel, This Location:
Lisa A. Conte, President
Gary L. Fischer, Vice-President

Sigma-Tau Pharmaceuticals, Inc.
200 Orchard Ridge Road
Gaithersburg, MD 20878

Telephone: (301) 948-1041

Key Personnel, This Location:
C. Kenneth Mehrling, Vice President

Smith & Nephew, Inc.
11775 Starkey Road
Largo, FL 34649

Telephone: (813) 399-9115

Key Personnel, This Location:
T. Furness, President
D. Southworth, Vice-President

SmithKline Beecham Healthcare, L.P.
P.O. Box 1467
Pittsburgh, PA 15230

Parent Company:
SmithKline Beecham, Inc.

Telephone: (412) 928-1050

SmithKline Beecham Pharmaceuticals
One Franklin Plaza, P.O. Box 7929
Philadelphia, PA 19101

Parent Company:
SmithKline Beecham, plc

Telephone: (215) 751-4000

Key Personnel, This Location:
Robert P. Bauman, President
John B. Ziegler, III, Vice President

Solvay Pharmaceuticals
901 Sawyer Road
Marietta, GA 30062

Parent Company:
Solvay America, Inc.

Telephone: (404) 578-9000

FAX: (404) 578-5586

Key Personnel, This Location:
James W. Warren, Jr., President
L. D. Muschek, Sr. Vice President

Somerset Pharmaceuticals, Inc.
777 South Harbour Island Boulevard, Suite 880
Tampa, FL 33602

Telephone: (813) 223-7677

E. R. Squibb & Sons, Inc
P.O. Box 4000
Princeton, NJ 08543

Telephone: (609) 252-4000

Parent Company:
Bristol-Myers Squibb Company

Star Pharmaceuticals, Inc.
1990 Northwest 44th Street
Pompano Beach, FL 33064

Telephone: (305) 971-9704

FAX: (305) 971-7718

Key Personnel, This Location:
Scott L. Davidson, President

Stiefel Laboratories, Inc.
255 Alhambra Circle, Suite 1000
Coral Gables, FL 33134

Telephone: (305) 443-3800

Key Personnel, This Location:
Werner K. Stiefel, President
Charles W. Stiefel, Exec. Vice President

Sterling Winthrop, Inc.
90 Park Avenue
New York, NY 10016

Telephone: (212) 907-2000

Stuart Pharmaceuticals
Wilmington, DE 19897

Telephone: (302) 886-3000

Parent Company:
Zeneca, Inc.

Summit Pharmaceuticals
556 Morris Avenue
Summit, NJ 07901

Telephone: (908) 277-5000

Parent Company:
Ciba-Geigy Corp.

Syntex Laboratories, Inc.
3401 Hillview Avenue
P.O. Box 10850
Palo Alto, CA 94303

Telephone: (415) 855-5050

Key Personnel, This Location:
James N. Wilson, President
Thomas L. Gutshall, Exec. Vice President

Parent Company:
Hoffmann-LaRoche, Inc.

TAP Pharmaceuticals, Inc.
2355 Waukegan Road
Deerfield, IL 60015

Telephone: (800) 622-2011

Taro Pharmaceuticals U.S.A., Inc.
Six Skyline Drive
Hawthorne, NY 10532

Telephone: (914) 345-9001 FAX: (914) 345-8727

Key Personnel, This Location:
Aaron Levitt, President
Paul Giallorenzo, Sr. Vice President

Taylor Pharmacal Company
1505 Wyckles Road
Decatur, IL 62525

Parent Company:
Akorn, Inc.

Telephone: (217) 428-1100

Key Personnel, This Location:
F.S. Taylor, President
B.J. Wattelet, Vice President

Thames Pharmacal Company, Inc.
2100 5th Avenue
Ronkonkoma, NY 11779

Telephone: (516) 737-1155

Key Personnel, This Location:
S.T. Blackamn, President
Richard McGee, Vice President

Thompson Medical Company, Inc.
222 Lakeview Avenue
West Palm Beach, FL 33401

Telephone: (800) 352-8466

Key Personnel, This Location:
Edward L. Steinberg, President

Triton Consumer Products, Inc.
561 West Golf Road
Arlington Heights, IL 60005

Telephone: (708) 228-7650

Tyson & Associates, Inc.
12832 Chadron Avenue
Hawthorne, CA 90250

Telephone: (213) 675-1080

UAD Labboratories
8339 Highway 18
Jackson, MS 39209

Telephone: (601) 372-7773

U.S. Pharmaceutical Corp.
2401C Mellon Court
Decatur, GA 30035

Parent Company:
Mead Johnson & Company

Telephone: (800) 330-3040

FAX: (404) 987-4806

Key Personnel, This Location:
S.L. Baker, President

The Upjohn Company
7000 Portage Road
Kalamazoo, MI 49001

Telephone: (616) 323-4000

Key Personnel, This Location:
L.S. Smith, President
D.R. Parfet, Exec. Vice President

Upsher-Smith Laboratories, Inc.
14905 23rd Avenue North
Minneapolis, MN 55447

Telephone: (800) 654-2299

Vitaline Formulas
722 Jefferson Avenue
Ashland, OR 97520

Telephone: (503) 482-9231 FAX: (503) 482-9112

Key Personnel, This Location:
J.D. Meese, Technical Director

Vitarine Pharmaceuticals, Inc.
227-15 North Conduit Avenue
Springfield Gardens, NY

Telephone: (718) 276-8600

Key Personnel, This Location:
Roger Jordan, President
Seymour Heydon, Exec. Vice President

Wakunaga Of America Company, Inc.
23501 Madero
Mission Viejo, CA 92691

Parent Company:
Wakunaga Pharmaceutical Co.

Telephone: (714) 855-2776

Wampole Laboratories
Half Acre Road, P.O. Box 1001
Cranbury, NJ 08512

Parent Company:
Carter-Wallace, Inc.

Telephone: (609) 655-6000

Key Personnel, This Location:
John Bridgen, President
B. Friedman, Vice President

Warner Chilcott Laboratories
201 Tabor Road
Morris Plains, NJ 07950

Parent Company:
Warner-Lambert Company

Telephone: (201) 540-2000

FAX: (201) 540-3283

Key Personnel, This Location:
Rick Wantiez

Warner-Lambert Company
201 Tabor Road
Morris Plains, NJ 07950

Telephone: (201) 540-2000

Key Personnel, This Location:
L.J.R. deVink, President
Joseph E. Smith, Exec. Vice President

Webcon Uriceutical
2 Constitution Way
Woburn, MA 02061

Parent Company:
Poly Medica Pharmaceuticals (USA), Inc.

Telephone: (617) 933-2020

FAX: (617) 933-7992

Key Personnel, This Location:
Peter M. Etzel

West Point Pharma
West Point, PA 19486

Telephone:(800) 637-2511

Parent Company:
Merck & Company, Inc.

Westwood-Squibb Pharmaceuticals, Inc.
100 Forest Avenue
Buffalo, NY 14213

Telephone: (716) 887-3400

Key Personnel, This Location:
J.B. Marsh, President
D.O. Smith, Vice President

Parent Company:
Bristol-Myers Squibb

Whitby Pharmaceuticals, Inc.
1211 Sherwood Avenue,
P.O. Box 85054
Richmond, VA 23220

Telephone: (804) 254-4400

Key Personnel, This Location:
Norman D. Schellenger, President
Geoffrey Alan, Vice President

Parent Company:
Ethyl Corp.

Whitehall Laboratories, Inc.
Five Giralda Farms
Madison, NJ 07940

Telephone: (800) 343-0856

Key Personnel, This Location:
Jean-Claude Leroux, President

Parent Company:
American Home Products Corp.

Wyeth-Ayerst Laboratories
P.O. Box 8299
Philadelphia, PA 19101

Parent Company:
American Home Products Corp.

Telephone: (215) 688-4400

Key Personnel, This Location:
Robert Essner, President
John V. Bucceri, Exec. Vice President

Zeneca Pharmaceuticals
Wilmington, DE 19897

Parent Company:
Zeneca, Inc.

Telephone: (302) 886-2231

Key Personnel, This Location:
Jeffrey S. Rudolf, Vice President

Zenith Laboratories
140 LeGrand Avenue
Northvale, NJ 07647

Telephone: (201) 767-1700

Key Personnel, This Location:
John H. Klein, President & CEO
Lenora C. Gavalas, Exec. Vice President

Zila Pharmaceuticals, Inc.
5227 North 7th Street
Phoenix, AZ 85014

Telephone: (602) 266-6700

Key Personnel, This Location:
Joseph Hines, President
Edward Pomerantz, Vice President

---

50. VENEZUELA

Bayer de Venezuela S.A.
Apartado Postal 6693
Caracas, D.F. 1010-A

Telephone: 2 905 2111 FAX: 2 905 2018

Key Personnel, This Location:
Claus Remy, President

Bristol-Myers de Venezuela S.A.
Apartado Postal 62412
Caracas, D.F. 1060-A

Telephone: 2 285 5911 FAX: 2 285 5993

Key Personnel, This Location:
Hector Perez-Moll, General Manager

Laboratorios Substantia C.A.
Apartado Postal 5828
Caracas, MIR. 1010-A

Telephone: 2 239 4411 FAX: 2 239 2980

Key Personnel, This Location:
Christian Weise, President

PART II

CONTRACT PACKAGERS OF PHARMACEUTICALS

ARGENTINA

Nerova S.A.
Alte Brown 2852
1752 Lomas del Mirador, Buenos Aires

Telephone: 1 655 1167

AUSTRALIA

Kolmar (Australia) Pty., Ltd.
45 Kings Road
Hornsby, New South Wales 2077

Telephone: 2 476 1900

FAX: 2 476 1431

Key Personnel, This Location:
Craig Hunter, Managing Director

BRAZIL

Aerojet Quimica Industrial
Rua Jura Sem, No. Quadra 537
Bairro Dois Irmaos
Duque de Caxais RJ 25230 090

Telephone: 21 776 1841

FAX: 21 776 1904

Key Personnel, This Location:
Dimitri Giokaris, President
Aspasia Giokaris, Executive Director

CANADA

Burroughs Wellcome
16751 Rte Trans-Canadian
Kirkland, Québec H9H 4J4

Telephone: (514) 694-8220

FAX: (514) 694 8201

Key Personnel, This Location:
Danielle Munn, Manager
Pierre Brazeau, Contract Work Supervisor

Custom Pharmaceuticals
977 Century Drive
Burlington, Ontario L7L 5J8

Telephone: (905) 639 5254

FAX: (905) 639-2506

Key Personnel, This Location:
Lori A. Mosca, Vice President

Delmar Chemicals, Inc.
9321 Airlie
Lasalle, Québec H8R 2B2

Telephone: (514) 366 7950

FAX: (514) 366 5665

Key Personnel, This Location:
Frank Schaub

Parke-Davis
2337 Parkedale Avenue
Brockville, Ontario K6V 5W5

Telephone: (613) 342-4436

FAX: (613) 342-6584

Key Personnel, This Location:
N. Reynolds, Mgr. Tech. Service
Ivan Zarenda, Plant Manager

Ivers-Lee Ltd.
31 Hansen Road South
Brampton, Ontario L6W 3H7

Telephone: (905) 451 5535

FAX: (905) 451-1255

Key Personnel, This Location:
Brian Munro, President

Therapex
11100 Colbert
Anjou, Québec H1J 2M9

Telephone: (514) 353 5820

FAX: (514) 351-3450

Key Personnel, This Location:
Monique Corbin, Vice President

Kolmar Of Canada
149 Victoria Street
Barrie, Ontario L4N 2J6

Telephone: (705) 728-5917

FAX: (705) 728 8262

Key Personnel, This Location:
G.A. Lee, Sr. Vice President

CYPRUS

Biochemie Rose, Ltd.
P.O. Box 3891
Nicosia

Telephone: 2 482 251

FAX: 2 482 271

FRANCE

Finoga
Route de Givors
38670 Chasse-sur-Rhône

Telephone: 72 49 19 60

FAX: 72 49 19 73

Sanofi Chimie
9 Rue du President Allende
94256 Gentilly, Cedex

Telephone: 1 40 73 40 73

FAX: 1 40 73 61 61

Laboratoire Gouze-Baudoux
11 Avenue Michel Chevalier
ZI des Bois
06130 Grasse

Telephone: 93 70 05 33

FAX: 93 70 05 27

INDIA

Accra Pac
917 Raheja Chambers
Nariman Point
Bombay 21

Telephone: 22 24 38 47

FAX: 22 20 43 66 1

Key Personnel, This Location:
Vikram Shah, Managing Director
Munna Shah, Director

Balsara Hygiene Products, Ltd.
43 Nagindas Master Road Fort
Bombay 400 001

Telephone: 22 26 54 23 6

FAX: 22 26 53 91 1

Key Personnel, This Location:
A.R. Balsara, Managing Director

Anant & Company
194 Arvind Chambers
Gauri Studio Compound
Western Express Highway
Andheri (E) 400 069 Bombay

Telephone: 22 837 50 59

FAX: 22 827 25 37

Vibro Pharma
Plot No. 14, Lelwani Ind. Estate
G.D. Ambekar Road
Wadala, Bombay 31

Telephone: 22 412 72 39

Key Personnel, This Location:
Rajendra K. Shah, Director

ISRAEL

Careline (Pharmagis), Ltd.
29 Lehi Street
51200 Bnei Brak

Telephone: 3 577 3666

FAX: 3 577 3878

ITALY

Laboratori Guidotti, S.p.A.
Via Trieste 4D
Pisa 56100

Telephone: 50 50 24 39

FAX: 50 40 25 0

Key Personnel, This Location:
Cosimo Bracci Torsi, President
P. Monelli, Director General

S.I.M.S. s.r.l
Località Filarone
Reggello FL 50066

Telephone: 55 86 30 51

FAX: 55 86 30 08

JAPAN

Nihon Kolmar Company
4-1 Fushimimachi 4-Chome, Chuo-ku
Osaka, 541

Telephone: 6 227 5261

FAX: 6 227 5264

Key Personnel, This Location:
Shigeru Kanzaki, President
Tomoji Kanzaki, Exec. Vice President

KOREA

Korea Kolmar Company
Samsung L B/D, 108-9, Samsung Dong
Kangnam Ku, Seoul

Telephone: 2 515 0150

FAX: 2 514 6479

Key Personnel, This Location:
Dong-Han Yoon, President

MALAYSIA

Inchcape Contract Manufacturing Company
Persiaran Perusahaan, Sek. 23
Shah Alam, Selangor 40000

Telephone: 3 541 6603

FAX: 3 541 6635

Key Personnel, This Location:
B.S. Lean, Managing Director

Kontrak Manufacturing
Services, Sdn. Bhd.
P.O. Box 7293, 40710
Shah Alam, Selangor

Telephone: 3 541 6600

FAX: 3 541 6635

Key Personnel, This Location:
Jessica Ng, Marketing Mgr.

MALTA

Biochemicals International
K2 Industrial Estate
Corradino, Paolo PLA 08

Telephone: 644 115

FAX: 677 211

SINGAPORE

Contract Manufacturing Services, Pte., Ltd.
2 Lorong Tukang Empat
Jurong Town
Singapore 2261

Telephone: 265 2222

FAX: 264 1554

Key Personnel, This Location:
Loh Kok Choy, General Manager

Sunward Chemical Industries Company Pte., Ltd.
11 Wan Lee Road, Jurong Town
Singapore 2262

Telephone: 262 6022

Key Personnel, This Location:
Niou Tien Kwao, Managing Director

Malaysia Chemist (Pte.), Ltd.
Blk. 104 #04-06, Boon Keng Road
Boon Keng
Industrial Establishment
Singapore 1233

Telephone: 298 5867

Key Personnel, This Location:
Goh Lim Koon, Managing Director

Union Chemical & Pharmaceutical (Pte.), Ltd.
113 Eunos Avenue 3
#06-06 Gordon Industrial Building
Singapore 1440

Telephone: 743 5402

Key Personnel, This Location:
Chan Szie Chang, Managing Dir.

SPAIN

Laboratorio Aldo-Union S.A.
Espulgues de Llobregat 08950
Barcelona

Telephone: 3 372 71 11

FAX: 3 371 61 98

Key Personnel, This Location:
J.S. Sanmartin, Technical Director

Laboratorios Margrina S.A.
C/ Varsovia 22
08026 Barcelona

Telephone: 3 51 74 11

FAX: 3 40 84 26 4

Uquifa
Mallorca 262
08008 Bacelona

Telephone: 3 487 94 77

FAX: 3 488 04 91

SWITZERLAND

ASM Aerosol Service
Industriestrasse 11
43 3 Möhlin

Telephone: 61 851 42 22 FAX: 61 851 50 94

Key Personnel, This Location:
E. Steinfels, Owner
M. Lütelzschwab, Marketing Director

THAILAND

Inchcape Manufacturing, Ltd.
21/7 Moo 6, Kookot
Lamlukka, Pathumtnani 12130

Telephone: 2 533 77 50

FAX: 2 531 73 70

Key Personnel, This Location:
Banchong Chittchang, Managing Director
S. Vongsarlyavanich, Commercial Director

Thai Milott Laboratories Co., Ltd.
84-55 Moo 2, Soi Tanasit
Teparak Road, Bang Pla
Bang Plee, Samut Prakarn

Telephone: 2 312 11 78

UNITED KINGDOM

Advanced Healthcare, Ltd.
Dukes Factory
Chiddingstone Causeway
Tonbridge, Kent TN11 8JU

Telephone: 892 870 500

FAX: 892 870 482

Boots Contract Manufacturing
Boots Company, plc
Nottingham NG2 3AA

Telephone: 602 593 116

FAX: 602 593 606

Berk Pharmaceuticals
Brampton Rd., Hampden Park
Eastbourne, E. Sussex BN22 9AG

Telephone: 323 501 111

FAX: 323 505 275

Brecon Pharmaceuticals, Ltd.
Forest Road
Hay on Wye, Herford

Telephone: 497 820 829

FAX: 497 820 050

Biotech (Suffolk), Ltd.
94 Hampstead Avenue
Mildenhall, Suffolk 1P28 7AS

Telephone: 638 713 100

FAX: 638 510 315

CCL Industries, Ltd.
Medical & Personal Care Division
Atkinsons Way, Foxhills Ind. Park
Scunthorpe
S. Humberside DN15 8QJ

Telephone: 724 282 220

FAX: 724 280 675

The Body Care Company, Ltd.
Unit 22 School St. Industrial Estate
Hazel Grove
Stockport, Cheshire SK7 4RA

Telephone: 61 419 9577

FAX: 61 419 9400

CCL Industries, Ltd.
Minipak Pharmaceuticals Division
5-9 Arkwright Road
Astmoor Industrial Estate
Runcorn, Cheshire WA7 1NU

Telephone: 9285 676 76

FAX: 9285 606 08

Chester Laboratories
Unit 11, First Avenue
Deeside Industrial Park
Deeside, Clwyd CH5 2NU

Telephone: 244 288 681

FAX: 244 280 933

Key Personnel, This Location:
James Shipp, Technical Director

Health & Diet , Ltd.
Stonecloth Road
Radcliffe, Manchester M26 9HE

Telephone: 204 707 420

FAX: 204 732 980

Creighton's Naturally, plc
Water Lane
Storrington, W. Sussex RH20 3DP

Telephone: 903 745 611

FAX: 903 745 986

JCL Pharmaceuticals, Ltd.
Limberline Road
Hilsea, Portsmouth PO3 5JS

Telephone: 705 678 100

FAX: 705 654 500

Fleet Laboratories, Ltd.
94 Rickmansworth Road
Watford, Herts WD1 7JJ

Telephone: 923 229 251

FAX: 923 220 728

Key Personnel, This Location:
Tony Hook
Kevin Duffy

Laleham Healt Care, Ltd.
Newman Lane
Alton, Hants GU34 2QR

Telephone: 420 890 07

FAX: 420 898 14

Hanworth Laboratories, Ltd.
The Grip, Hadstock Road
Linton, Cambs

Telephone: 223 892 217

FAX: 223 893 843

Lorco Hygiene
33 Croft Street
Clayton, Manchester M11 4RQ

Telephone: 61 223 8261

FAX: 61 230 7803

Lorien Laboratories, Ltd.
Lorien House, Sunderland Road
Sandy, Beds SG19 1RB

Telephone: 767 291 990

FAX: 767 291 908

William Ransom & Son, plc
Hitchin
Hertfordshire SG5 1LY

Telephone: 462 437 615

FAX: 462 420 528

Medi-Pack, Ltd.
101 Hessel Road, P.O. Box 20
Hull HU3 2BW

Telephone: 482 251 81

FAX: 482 283 26

Robinson Healthcare
Hipper House
Chesterfield S40 1YF

Telephone: 246 220 022

FAX: 246 208 164

Pharmaco UK, Ltd.
140 Cambridge Science Park
Milton Road
Cambridge CB4 4GF

Telephone: 223 425 777

FAX: 223 423 838

Sauflon Pharmaceuticals, Ltd.
6/7 Mace Industrial Estate
Ashford, Kent TN24 8EP

Telephone: 233 646 610

FAX: 233 616 614

Pharmasol, Ltd.
North Way
Walworth Industrial Estate
Andover, Hants SP10 5AZ

Telephone: 264 363 117

FAX: 264 332 223

Smith & Nephew Consumer Products, Ltd.
Alum Rock Road
Birmingham B8 3DZ

Telephone: 21 327 4750

FAX: 21 327 6172

Wrafton Laboratories
Wrafton, Brounton
North Devon

Telephone: 271 815 815

FAX: 271 815 053

UNITED STATES

Abbott Laboratories,
Pharmaceutical Products
Dept.-53T, 1401 Sheridan Road
North Chicago, IL 60064

Telephone: (708) 937-1009

FAX: (708) 938-2875

Key Personnel, This Location:
A. Rosenston

Accupac, Inc.
Wambold Road At
Industrial Boulevard, P.O. Box 200
Mainland, PA 19451

Telephone: (215) 256-7000

FAX: (215) 256-7009

Key Personnel, This Location:
John J. Braithwaite, President
A. Bruce Hech, Exec. Vice Pres.

Accra Pac Group
2370 Middlebury Street
P.O. Box 878
Elkhart, IN 46515

Telephone: (219) 295-0000

Key Personnel, This Location:
Satish Shah, President
Kenneth Wanner, Sales Manager

Aerosol Services Company, Inc.
425 South Ninth Street
City of Industry, CA 91746

Telephone: (818) 968-8531

FAX: (818) 330-9464

Key Personnel, This Location:
Walter Lim, President

Anderson Packaging
4545 Assembly Drive
Rockford, IL 61109

Telephone: (815) 397-6091

FAX: (815) 397-6094

Key Personnel, This Location:
Larry Beckmann, President
Edward Hancock, Vice President

Bell-More Labs, Inc.
4030 Gill Avenue
Hampstead, MD 21074

Telephone: (410) 239-7554

Banner Pharmcaps
20730 Dearborn Street
Chatsworth, CA 91311

Telephone: (818) 341-3060

FAX: (818) 341-1347

Key Personnel, This Location:
Stephen Fischer, President
Brenda Peterson, Sales Manager

Ben Venue Laboratories
270 Northfield Road
Bedford, OH 44146

Telephone: (216) 232-3320

FAX: (216) 232-2772

Key Personnel, This Location:
Thomas R Russillio, President

Baxter Contract Manufacturing
One Baxter Parkway
Deerfield, IL 60015

Telephone: (708) 270-4833

FAX: (708) 270-3765

Key Personnel, This Location:
Paul Cummings Sr. Vice President
Maryellen Zibell

CCL Custom Manufacturing
6133 North River Road, Suite 800
Rosemont, IL 60018

Telephone: (708) 823-0060

FAX: (708) 823-0061

Key Personnel, This Location:
Donald Lang, President
Richard Clark, Vice President

Chemical Science Laboratories
3605 West 96th Terrace
Lexena, KS 66215

Telephone: (913) 541-0525

FAX: (913) 888-3582

Copper Packaging
P.O. Box 248
Elkhart, IN 46515

Telephone: (219) 264-4195

Key Personnel, This Location:
Louis Minelli, President

Conair Packaging Company
1 Cummings Point Road
Stamford, CT 06904

Telephone: (203) 351-9100

FAX: (203) 351-9180

Key Personnel, This Location:
Lee Rizzuto, Sr., President
Dennis Rizzuto, Vice President

Corwood Laboratories
55 Adams Avenue
Hauppauge, NY 11788

Telephone: (516) 273-7373

FAX: (516) 273-7465

Key Personnel, This Location:
Irwin Thaler, President
Laurence H. Levine, Sales Mgr.

Continental Laboratories
912 South State Street
Madrid, IA 50156

Telephone: (515) 795-2000

FAX: (515) 795-2954

Key Personnel, This Location:
David A. Bequeaith, President
Charles Insalaco, Vice President

Cosmetic Group
3216 Gillham Plaza
Kansas City, MO 64109

Telephone: (816) 531-2685

FAX: (816) 531-6979

Key Personnel, This Location:
Grant Grissom, Vice President

Cosmetic Group USA
11312 Penrose Street
Sun Valley, CA 91352

Telephone: (818) 767-2889

FAX: (818) 767-4062

Key Personnel, This Location:
Alfred E. Booth, Jr., President
Judy Zagarelli, Vice President

DPT Laboratories
307 E. Josephine Street
San Antonio, TX 78215

Telephone: (210) 223-3281

FAX: (210) 224-6505

Key Personnel, This Location:
John Feik, President

C.P. Packaging
1075 Cranbury South River Road
Jamesburg, NJ 08831

Telephone: (609) 655-4880

FAX: (609) 655-5718

Key Personnel, This Location:
J. O'Meara
R. O'Connell

Dimentional Merchandising, Inc.
120 Dorsa Avenue
Livingston, NJ 07039

Telephone: (201) 535-2900

FAX: (201) 535-0909

Key Personnel, This Location:
Eugene R. Sylva, President
Robert A. Riker, Vice President

Creative Products Of Rossville
One Creative Way, P.O. Box 125
Rossville, IL 60963

Telephone: (217) 758-6784

FAX: (217) 741-6784

Key Personnel, This Location:
Daniel E. Clapp, President
C. H. Clapp, Jr., Vice President

EMS Contract Packaging
321 Morris Road, P.O. Box 162
Fort Washington, PA, 19034

Telephone: (215) 646-1477

FAX: (215) 646-8931

Key Personnel, This Location:
Stacie Kirsch, President
James Magyar, Vice President

Environmental Packaging Group
5811 Marlborough Drive
Cincinnati, OH 45230

Telephone: (513) 232-8890

FAX: (513) 232-0856

Key Personnel, This Location:
Dennis O. Howland, President

Gordon Laboratories
400 North Nash Street
El Segundo, CA 90245

Telephone: (213) 772-0083

Key Personnel, This Location:
Morgan D. Gaffney, President
William J. McDermott, Plant Mgr.

Fluid Packaging Company, Inc.
800 Airport Road
Lakewood, NJ 08701

Telephone: (908) 367-1000

FAX: (908) 367-2114

Key Personnel, This Location
Edmund Bennett, President
John Bennett, Vice President

Hydrotech Laboratories
95 East Bradrock Drive
Des Plaines, IL 60018

Telephone: (708) 296 9545

Key Personnel, This Location:
Kurt M. Johnson, President
Laura Jenson, R&D Manager

Goldschmidt Cosmetics
76 Bridge Street
Islandia, NY 11722

Telephone: (516) 234-5900

FAX: (516) 323-1203

Key Personnel, This Location:
Joseph Goldschmidt, CEO
Les Goldschmidt, President

J&B Systems
5055 State Route 276
Batavia, OH 45103

Telephone: (513) 732-2000

Key Personnel, This Location:
Jerrilyn Kearney, President
William Kearney, Vice President

Kasper, Inc.
200 Corporate Drive
Mahwah, NJ 07430

Telephone: (201) 529-1600

Key Personnel, This Location:
Frank O'Neill, President

Libby Laboratories
1700 Sixth Street
Berkeley, CA 94710

Telephone: (510) 527-5400

FAX: (510) 527-8687

Key Personnel, This Location:
Susan Libby, President

Kolmar Laboratories
King Street
Port Jervis, NY

Telephone: (914) 856-5311

FAX: (914) 856-2203

Key Personnel, This Location:
Chris Denney, President
James Cooper, Sr. Vice President

Marvell Pharmacal & Packaging Company
490 Route 46 East
Faifield, NJ 07004

Telephone: (201) 882-9339

Key Personnel, This Location:
Richard Teich, President
Scott Teich, Vice President

Alvin Last, Inc.
19 Babcock Place
Yonkers, NY 10701

Telephone: (914) 376-1000

FAX: (914) 376-1001

Key Personnel, This Location:
Alvin Last, President
Bruce Last, Vice President

Medicia Pharmaceutical
RD 3 Homestead Road
Belle Mead, NJ 08502

Telephone: (908) 359-6131

Key Personnel, This Location:
J. Paterno, President
R.M. Jose, Vice President

Nice-Pak Products
Two Nice-Pak Park
Orangeburg, NY 10962

Telephone: (914) 365-1700

FAX: (914) 365-1717

Key Personnel, This Location:
Robert P. Julius, President
Jonathan Kupperman, Director

Paco Pharmaceutical Services
1200 Paco Way
Lakewood, NJ 08701

Telephone: (908) 367-9000

FAX: (908) 364-5266

Key Personnel, This Location:
James W. Haney, President
James C. Rhodes, Vice President

Noveau Labs
One Anderson Avenue
Moonachie, NJ 07074

Telephone: (201) 438-3200

FAX: (201) 460-4797

Key Personnel, This Location:
John Kressaty, Vice President

Palm Beach Beauty Products
950 Xenia Avenue
Minneapolis, MN 55416

Telephone: (800) 326-7256

FAX: (612) 546-2548

Key Personnel, This Location:
Ben Kaitz, President
Richard Kaitz, Exec. Vice President

Packaging Coordinators
K Street & Erie Avenue
Philadelphia, PA 19124

Telephone: (215) 537-8100

FAX: (215) 743-4818

Key Personnel, This Location:
Daniel Gerner, President
Paul Alvater, Sr. Vice President

Parke-Davis
870 Parkdale Road
Rochester, MN 48307

Telephone: (810) 656-5305

Key Personnel, This Location:
John Schuler

Pharmasol Corp.
One Norfolk Avenue
South Easton, MA 02375

Telephone: (508) 238-8501

FAX: (508) 238-0105

Key Personnel, This Location:
Scott A. McCaig, President
Gerald Strickland, Sales Director

Qualis, Inc.
4600 Park Avenue
Des Moines, IA 50321

Telephone: (800) 334 4514

FAX: (515) 282-1417

Key Personnel, This Location:
Richard Dowling, President
Michael S. Hatten, Vice President

Pharma-Turm, Inc.
Heston Hall-Carriage House
1790 Yardley-Langhorne Road
Yardley, PA 19067

Telephone: (215) 321-6930

FAX: (215) 321-6932

Key Personnel, This Location:
Jeff Turns

Raani Corp.
5401 West 65th Street
Bedford Park, IL 60638

Telephone: (708) 496-8025

FAX: (708) 496-8906

Key Personnel, This Location:
R.A. Chaudary, President
Eugene M. Frank, Vice President

Private Label Cosmetics
20-10 Maple Avenue
Fair Lawn, NJ 07410

Telephone: (201) 423-1515

FAX: (201) 423-3786

Key Personnel, This Location:
Michael J. Assante, President
Santo Spagnuollo, Vice President

M.J. Rapoport & Company, Inc.
513 Central Avenue
Highland Park, IL 60035

Telephone: (708) 926-0101

FAX: (708) 926-0139

Key Personnel, This Location:
Michael J. Rapoport, President

Rempak Industries
1600 St. Georges Avenue
Rahway, NJ 07065

Telephone: (908) 382-2400

FAX: (908) 382-2002

Key Personnel, This Location:
Edward G. Thau
Eugene Cohen

San Mar Laboratories
4 Warehouse Lane
Elmsford, NY 10523

Telephone: (914) 592-3130

Key Personnel, This Location:
Marvin Berkrot, President
Irwin Silverberg, Vice President

Roussel Uclaf Corp.
95 Chestnut Ridge Road
Montvale, NJ 07645

Telephone: (201) 307-9700

FAX: (201) 307-3281

Key Personnel, This Location:
T. Driscoll, President

Schering-Plough
Third-Party Business
Kenilworth, NJ 07033

Telephone: (908) 629-3200

FAX: (908) 629-3164

Ruspak Corp.
P.O. Box 29, Manhatten Street
Lyons, NY 14489

Telephone: (315) 946-9777

FAX: (315) 946-9525

Key Personnel, This Location:
Donald Bickle, President
Paul Mierzwa, Sales Manager

Sorenco, Inc.
2323 South West Temple
Salt Lake City, UT 84115

Telephone: (801) 481-7325

FAX: (801) 467-9024

Key Personnel, This Location:
Robert Tingey, President
Robert Kaelin, Vice President

Thames Phamacal
2100 Fifth Avenue
Ronkonkoma, NY 11779

Telephone: (516) 737-1155

Key Personnel, This Location:
Eric Stern, Vice President
Paula Fierro, Sales Director

VEE PAK
5321 Dansher Road
Countryside, IL 60525

Telephone: (708) 482-8881

FAX: (708) 482-3343

Key Personnel, This Location:
Ralph S. Vennetti, President
Paul M. Morrison, Vice President

Unette Corp.
26 Eastmans Road
Parsippany, NJ 07054

Telephone: (201) 884-0010

FAX: (201) 884-0412

Key Personnel, This Location:
George F. Wohlgemuth, President

Virbac, Inc.
3200 Meacham Boulevard
Fort Worth, TX 76161

Telephone: (800) 338-3659

FAX: (817) 831-8327

Key Personnel, This Location:
Thomas Adcock, Vice President

Valjean
1785 South Patrick Drive
Indian Harbour Beach, FL 32937

Telephone: (407) 773-2454

FAX: (407) 773-0988

Key Personnel, This Location:
Richard T. Ferry, President
